Rejectionist Philosophy from
Buddhism to Benatar

First Edition Design Publishing

Sarasota, Florida

Anti-Natalism:
Rejectionist Philosophy from Buddhism to Benatar
Copyright ©2014 Ken Coates

ISBN 978-1506-902-40-1 PAPERBACK
ISBN 978-1622-875-70-2 EBOOK

July 2016

Published and Distributed by
First Edition Design Publishing, Inc.
P.O. Box 20217, Sarasota, FL 34276-3217
www.firsteditiondesignpublishing.com

Professor David Benatar's example and encouragement have been important in writing this book. He also read and commented on parts of it. My grateful thanks to him. I, however, remain solely responsible for its contents.

Foreword

By David Benatar

Sadly, I never met the late Professor Ramesh Mishra. We did, however, communicate via email. He first contacted me in December 2009 after reading my book, *Better Never to Have Been*. In subsequent emails he reported ill health, but in February 2014 he sent me a draft of the book for which I am now writing a foreword. I read parts of it and was very impressed by what I read.

It is a wide-ranging book on the theme of anti-natalism – or, to use Professor Mishra's preferred term, "rejectionist philosophy". Anti-natalism is the view that coming into existence is a serious harm. We ought not to inflict it on others by procreating, and we should regret it even when we cannot prevent it. Professor Mishra has a clear, accessible style of writing and the book is also impressive in the breadth of material covered – not only philosophical but also religious and literary expressions of the anti-natalist idea. While Professor Mishra was not a philosopher by training he wrote more competently about the philosophical issues in this book than some philosophers.

Born in Calcutta, India, Professor Mishra trained as an aircraft engineer and was drawn into trade unionism. This developed his interest in social justice. He abandoned engineering, moved to England, where he studied sociology and social policy, and received a PhD from the London School of Economics. He taught in the UK and Canada. At the time of his communication with me he was

i

Emeritus Professor of Social Policy and a Senior Scholar at York University in Toronto.

Professor Mishra originally published his manuscript as an e-book under the nom-de-plume "Ken Coates". At first he was keen to keep his actual identity secret and I respected this wish when an admiring reader asked whether I knew the author. However, after another period of serious illness and hospitalization, Professor Mishra wrote to give me permission to reveal his identity. He also indicated his intention to publish a paperback version of the book.

In April 2015 I received the sad news from his wife, Mrs. Heather Mishra, that her husband had passed away. There are millions of people with whom our lives do not overlap. Some die before we are brought into existence. Others will begin their lives once we have died. Of all the people who share the planet with us for some period, our lives intersect with only a few. Often the intersection is fleeting. We pass them in the street, sit next to them on a plane or train, meet them at some gathering or teach them for a semester. Sometimes, especially in our contemporary world made smaller by telecommunications, we never meet in person but we do connect electronically as Professor Mishra and I did in the final years of his life. I am grateful that we did.

Before Professor Mishra died, he had asked his wife to arrange for me to write a few words for the paperback version of *Anti-Natalism: Rejectionist Philosophy from Buddhism to Benatar,* that he had been planning. I am honoured to do so and to recommend his final book to prospective readers.

Anti-Natalism:

Rejectionist Philosophy from Buddhism to Benatar

Ken Coates

Table of Contents

Preface

The last few decades seem to have begun what has been called 'the childless revolution'. In economically developed countries more and more people are choosing not to have children. The causes of this 'revolution' are many. One of these is the belief that to create a new life is to subject someone unnecessarily, and without their consent, to life's many sufferings including death. This belief and its underlying philosophy is known as anti-natalism. There has been a recent resurgence of this philosophy, with David Benatar's book <u>Better Never To Have Been</u> (2006) as a major catalyst. Anti-natalism can be seen as part of a larger philosophy, described here as Rejectionism, which finds existence –   directly or indirectly, i.e. as procreation - deeply problematic and unacceptable.

The book traces the development of this philosophy from its ancient religious roots in Hinduism (Moksha) and Buddhism (Nirvana) to its most modern articulation by the South African philosopher, David Benatar (2006). It examines the contribution to rejectionist thought by Arthur Schopenhauer and Eduard von Hartmann in the 19th century and Peter Wessel Zapffe, a little known Norwegian thinker, in the 20th century and most recently by Benatar. In part the unfolding of this philosophy over the centuries is the story of a transition from a religious to a secular – at first metaphysical, and later to a positivist approach in the form of anti-natalism. Zapffe and Benatar represent the anti-natal approach most clearly.

The book also devotes a chapter to the literary expression of rejectionist philosophy in the works of Samuel Beckett and Jean Paul

1

Sartre. In sum, far from being an esoteric doctrine rejectionism has been a major presence in human history straddling all three major cultural forms – religious, philosophical, and literary.

The book argues that in developed countries where procreation is a choice, natalism or having children (Acceptance) is as much a philosophical stance in need of justification as its opposite, i.e. anti-natalism (Rejection). Secondly, the recent advance of anti-natal practice and the possibility of its further progress owe a great deal to three major developments: secularization, the liberalization of social attitudes, and technological advances (contraception). Anti-natal attitudes and practice should therefore be seen as a part of 'progress' in that these developments are widening our choice of lifestyles and attitudes to existence. Thirdly, and it follows, that anti-natalism needs to be taken seriously and considered as a legitimate worldview of a modern, secular civilization. Recent critique of anti-natalism has tended to treat it as a deviant or esoteric, if not a bizarre, viewpoint, restricted to a fringe or counter-culture. This is to misunderstand or misrepresent anti-natalism, and one of the objectives of this book is to situate current anti-natalist thought in its historical and philosophical perspective. Finally, it is argued that in order to further the development of anti-natalism it needs to be institutionalized as a form of rational 'philosophy of life' and more attention needs to be paid to the problems and prospect of putting this philosophy into practice.

Introduction

Human beings are the only creatures conscious of their own existence. Other living beings do not *know* that they exist. They cannot help going on living - reproducing and continuing the species - as programmed by nature. Humans alone have the capacity to interrogate their own existence. Since the dawn of consciousness human beings have found themselves confronting an existence they did not choose and which puts them through a great deal of pain and suffering – physical and mental - leading eventually to death.

To make life with all its multifarious evils acceptable and meaningful humans have invented religion, a supernatural system of beliefs, which, among other things, seeks to justify and legitimize existence. Yet even religions have not found it easy to endorse life with all its evils – man-made and natural – and have sought ways of emancipation from it[1]. For example Hinduism and Buddhism, with their concepts of Moksha and Nirvana respectively (Koller 1982, 67; Snelling 1998, 54-5) point a way of transcending the phenomenal world with its recurring cycle of births and deaths. In addition, secular philosophies which consider existence to be a 'bad' rather than a 'good' have their own views about the evil of existence and the way out.

Modern - mid-20th century onwards - secular philosophies see anti-natalism, i.e. refraining from procreation, as the way to liberation[2]. Besides expressing compassion for the unborn the decision not to reproduce is also a way of saying no to human existence. What these religious and secular philosophies have in common is the view that life in general and human life in particular is inherently flawed and that overcoming it would be a 'good' thing. While other creatures cannot escape their bondage to nature human beings can. They have the capacity to free themselves from the yoke of nature and to end their entrapment. And so they should. Broadly, the religious approach is based on freeing oneself from the will-to-live and the bondage to worldly desires whereas modern secular philosophies see anti-natalism as the key to emancipation.

But surely the prescription of anti-natalism is counter-intuitive? Our instincts make us want to live and to reproduce. The sex drive is one of the strongest physical urges and, in the absence of contraception, results naturally in reproduction. True, as anti-natalists remind us, the coming of contraception has sundered the natural bond between coitus and conception. Celibacy is no longer necessary in order to prevent reproduction. The sex urge need not be denied to avoid conception. And as far as an 'instinct' to reproduce is concerned this remains a somewhat dubious proposition at least as far as humans are concerned.

However, a more important objection against these philosophies is that they are unduly pessimistic and one-sided. They seem to turn a blind eye to all that is positive about life. For if there is much pain and suffering there is also pleasure, joy, love, beauty, creativity and the like. In short, life comes as a package deal, with good and evil inextricably mixed together. How can one separate them? Why dwell on the negativities of existence alone forgetting the other side? These are weighty arguments and they have to be taken seriously. They raise important philosophical issues which will be considered later

(Chapters 2, 3, 5 and 6). At this stage we would like to spell out the rationale for looking at these anti-existential viewpoints.

Let us start by noting that from time immemorial literary and philosophical writings have given expression to the feeling of outrage at the evil of existence. Hamlet's famous soliloquy is perhaps the best-known example. The most universally recognized symbol in Western civilization, the Cross carried by Christ, is a powerful message of life as a burden borne by man at the behest of God. In short, the viewpoint of life as evil has been a part of human consciousness. Anti-natalism too has its expression in literature such as Hamlet's admonition to Ophelia: "Why woulds't thou be a breeder of sinners? I am myself indifferent honest, but yet I could accuse me of such things that it were better my mother had not borne me .....What should such fellows as I do crawling between earth and heaven?...Go thy ways to a nunnery".

However despite a long history of literary allusions to the ills of existence, systematic philosophies, especially secular ones, which argue the case against existence are few and far between. They only date back to the 19th century, with Schopenhauer and to a lesser extent Eduard von Hartmann as the outstanding figures. And although Schopenhauer's thought includes a strong expression of anti-natalism - both as compassion to the unborn and as refusal to prolong the misery of existence - a philosophical treatise arguing the case for anti-natalism has appeared for the first time only recently, just a few years ago (Benatar 2006). In short, anti-existential philosophy and more especially the philosophy of anti-natalism constitutes a relatively recent and peripheral body of thought. It deserves to be known and discussed more widely. However what is common to the religious and secular philosophies presented in this book is their rejection of existence and the search for a way out. We therefore feel justified in using the term 'rejectionism' to indicate the genre of these philosophies and the chief characteristic of their world view[3].

A second reason for paying attention to rejectionist thinking is that as conscious beings we not only have the *capacity* to evaluate human existence we have a *duty* to do so. In order to do this and to make *authentic* choices concerning existence we need to be fully aware of our situation. The anti-existential perspective helps to deepen our awareness. For example, the decision whether to procreate or not is one of the most significant moral and metaphysical decisions we have to make in our lives. Quite recently Christine Overall (2012), a philosopher who is not an anti-natalist, has drawn attention to the moral issues involved in the decision to have a child, an issue she discusses quite comprehensively. Clearly, in making this decision we need to take anti-natalist arguments on board.

Anti-natalists, e.g. Benatar (2006), Hayry (2004), Srivastava (2006), argue that bringing someone into the world who has not asked to be born, to thrust life upon them and to put them through the painful business of living constitutes an immoral act. Thus children come into the world literally as someone else's creature and we can say that their life is founded in unfreedom. They are conscripts to life. Moreover, they are often considered as a means to an end, i.e. we produce children to serve *our* needs and interests, to entertain us, to pass *our* genes on, to ensure *our* biological continuity, to look after us in old age etc. Apart from its dubious *morality*, procreation also raises metaphysical issues. It amounts to endorsing existence and makes us indirectly complicit in all the evils that existence entails. Thus we need to be fully aware of the metaphysical choice and responsibility involved in the act of procreation.

These are only some of the philosophical issues surrounding procreation and we need to consider anti-natal arguments seriously. This is particularly important since sexual intercourse and reproduction comes 'naturally' to us – instincts, social conventions and religious teachings all conspire to incline us that way. Consider, for example, the age-old idea that women have a maternal instinct that craves satisfaction. it is only recently that this has been found to be a myth. Motherhood as the essential destiny of women has turned out to be little more than a natalist ideology propagated historically by patriarchal institutions. Millions of women, especially in developed countries, are choosing not to have children with apparently no instinctual urge to reproduce. Nonetheless voluntary childlessness, especially on the part of the married, remains taboo in Western 'advanced' societies. It is still considered as a form of deviant behavior (Defago 2005; Basten 2009; Overall 2012). Given the strength of conventional wisdom and the status quo it is important to pay attention to dissenting viewpoints concerning procreation.

Finally the importance of rejectionist thought lies in its open articulation of value judgments concerning life. This goes against the dominant view of philosophy, especially in the English-speaking world, which became established in early 20th century, viz. that it is not the business of philosophy to make value judgments, since they

7

amount to little more than stating the personal preferences of the philosopher. Logical positivism, and linguistic analysis, two of the three major currents of philosophical thought in the 20th century, held this view strongly. The third, viz. existentialism, differs radically in many ways from the other two. It is centrally concerned with the nature and problems of human existence. But unlike the theistic existentialism of Kierkegaard and others, non-theistic approaches, notably those of Heidegger and Sartre also stay away from making value judgments. Rather they emphasize individual freedom to decide and choose. Both thinkers stress the importance of *authentic* choice, i.e. a choice freely arrived at, in full awareness of one's situation as well as of the wider implications of one's choice (Sartre 1948; Watts 2001, 34-5, 56-60). And although Heidegger provides systematic and insightful analysis of man's being-in-the-world, e.g. the contingent nature of our birth, our awareness of temporality and the finite nature of our existence, he has little to say about procreation and most certainly nothing by way of an overall judgment about the nature of human existence (Magee 1978, 82-3, 92). Sartre, on the other hand, expresses strong rejectionist, including anti-natal views in some of his literary writings, notably in his landmark novel <u>Nausea</u> published in 1938. But it is Roquentin, the hero of the novel, who expresses these views and they cannot be equated with those of the author. Although Sartre's own biography seems to corroborate his anti-natal stance there is very little in his philosophical writings on procreation. Moreover Sartre's philosophy, not unlike that of Heidegger, makes no particular value judgments concerning existence. The idea is that each individual must decide for herself in full awareness of her situation

(Sartre 1948). We should also note that Sartre's celebrated lecture on existentialism (Sartre 1948) claims that it is a form of humanism implying that it affirms or at least accepts human existence as a given. In sum we could say that atheistic existentialism, as it developed in the 20th century, not only steers clear of value judgments about existence but also fails to consider the philosophical significance and implications of procreation.

True, there is by now a large body of philosophical literature concerned with aspects of procreation, e.g. the right and duty to have children, abortion, in-vitro fertilization, surrogate motherhood, and many other bio-ethical issues (Overall 2012, 12; Singer 1993). But much of it is an ad hoc treatment of these issues on moral grounds and quite unrelated to the broader philosophy of existence, including that of procreation. (Benatar 2004, especially Introduction; Overall 2012, 12-3). As we shall see later (Ch.3) Benatar's work may be seen as a step towards bridging this gap. Moreover much of the writings of the philosophers relevant to these issues tend to be 'pro-life' rather than 'pro-choice' and from this viewpoint too rejectionist philosophy deserves attention.

Besides religious and philosophical approaches we also look at modern literary perspectives with a rejectionist viewpoint. The work of Samuel Beckett and Jean-Paul Sartre may be considered as key exemplars of this viewpoint. Beckett is undoubtedly the anti-existential writer par excellence. In this regard Beckett is to literature what Schopenhauer is to philosophy. As for Sartre, his early novel Nausea is a landmark literary presentation of the contingency and superfluity of existence. In this early work his stance is clearly against existence. Few modern literary works are comparable in their expression of anti-existential viewpoint to those of Beckett and Sartre.

The Inclusion of literary perspectives besides the religious and philosophical broadens the anti-existential canvas. It underlines the universality or generality of the phenomenon which finds its expression in all the major cultural forms – religious, philosophical and artistic - concerned with the human condition. Of course literature is only one of the artistic modes of expression but it is the one that speaks most directly and intelligibly about the experience of existence. True, unlike the religious or philosophical modes it is not meant to provide either a diagnosis or a solution of the problem. But in giving artistic expression to the malaise of existence it adds a qualitatively different dimension to our consciousness of the problem. Major exemplars of all three modes of anti-existential expression – religious, philosophical and literary – are examined in this book.

The book is organized as follows. Chapter 1 is about religious perspectives. It explores the Hindu and Buddhist views of existence paying particular attention to the concepts of moksha and nirvana and the means and ends of liberation articulated by these religions. We note the spiritual and mystical nature of their approach to transcending existence.

The next chapter takes a critical look at 19th century secular philosophies concerned with existence and its transcendence. Schopenhauer is undoubtedly the first major Western philosopher to articulate the view that existence in general, and human existence in particular, is intrinsically and irremediably evil. Hence his reputation as the 'pessimistic' philosopher par excellence. However his philosophy has more to do with elaborating the nature and source of gratuitous suffering which life invariably entails than with finding a way out. His key concept of liberation is a state of 'willlessness' or the abdication of the will-to-live, arrived at voluntarily or otherwise. It resembles the Buddhist notion of nirvana, a similarity acknowledged by Schopenhauer. We consider next Eduard von Hartmann whose magnum Opus *The Philosophy of the Unconscious* (1869) seeks to combine an evolutionary perspective on human history with Schopenhauerian insights into the pain and suffering of existence. He foresaw a time when reason in man will prevail against his unconscious will-to-live resulting in collective worldwide action to bring an end to existence. The evolutionary historical perspective, the idea of an inherent conflict between will and reason, and that of a collective solution to the problem of existence are ideas which make Hartmann's work interesting and significant. He seems to have been unjustifiably neglected in the history of rejectionist thought. In Chapter 3 we move forward to the 20th century and beyond. We outline briefly the thinking of P.W. Zapffe(1899-1990), a little known Norwegian writer and philosopher, who was an early advocate of non-procreation as the solution to the irremediable contradictions of human existence. Finally we examine the work of a contemporary philosopher, David Benatar, whose book *Better Never To Have Been* *(*2006) is a seminal contribution to anti-existential thought. Benatar may be described as Schopenhauerian in his evaluation of human existence but his prime focus is on anti-natalism. He argues that it is our *duty* not to procreate since bringing any life into existence involves inflicting harm, i.e. pain and suffering including death. If each individual refrains from procreation a gradual phasing out of human existence becomes possible. Thus while Schopenhauer espouses a path to liberation which echoes Buddhist nirvana, Hartmann, Zapffe and Benatar suggest very different modes of liberation.

Chapter 4 presents the literary perspectives of Samuel Beckett and Jean-Paul Sartre on existence. Chapter 5 concludes by arguing that the philosophy of rejectionism, especially in its anti-natalist version, must be seen as one of the modern world-views that is here to stay and is likely to become more influential. It may contribute to increasing rejection of procreation based on moral and metaphysical considerations. The book ends (Chapter 6) with a set of hypothetical FAQs (Frequently Asked Questions and Answers) on rejectionism which provides the reader with an overview of this approach.

Endnote - Introduction

1. Given the wide variety, complexity, and ambiguity of many religions and religious belief systems any generalization must involve drastic simplification. What needs to be emphasized however in this context is the dualism of most world religions. On the one hand there is recognition of the 'evils' of this world and its denunciation and devaluation. This goes along with the promise of a better world to come in the hereafter for the faithful. On the other hand there is also an emphasis on human life as fundamentally good and to be accepted or even celebrated. The injunction to create progeny is a part of this endorsement of life. In several world religions such as Hinduism and Buddhism there is a distinction between mass and elite religiosity. It is the latter that is primarily concerned with world-rejection, asceticism and emancipation from the wheel of life. Some of these issues are discussed later(Chs. 1, 5 and 6).

2. Secular philosophies which consider existence, especially human existence, as seriously flawed are few and far between. Among these, philosophies that are explicitly anti-natalist include those of Zapffe and Benatar (see Ch. 3). Schopenhauer's position, often described as anti-natalist, is somewhat ambiguous in that although he is against procreation it is renunciation of the will that he considers as the path to liberation (see Ch. 2).

3.   These anti-existential perspectives may be termed 'rejectionist' in so far as they reject existence and seek to transcend it. 'Anti-natalism' means being opposed to procreation on philosophical grounds and implies the rejection of existence. But not all rejectionist perspectives are opposed to procreation. Hinduism, Buddhism and the philosophy of Hartmann are prime examples.

Chapter 1

Religious Perspectives

Among world religions Hinduism and Buddhism stand out in their strongly negative view of existence. Liberating human beings from their bondage to earthly existence has been their chief concern (Koller 1982, Chs. 4, 5, 7; Herman 1991, 114-18). Liberation is conceptualized as moksha or release in Hinduism and nirvana or extinction in Buddhism. Both these religions originated in India and have a great deal in common in their view of human existence and in their approach to transcending the evil of existence. However despite many similarities there are also important differences between the two and it is important to examine them separately. Hinduism is the older of the two and, directly or indirectly, has had a great deal of influence on Buddhism. Thus both chronologically and logically it makes sense to start with Hinduism. The main questions we ask of these belief systems are the following. How do they perceive human existence ? Wherein, according to them, lies the 'evil' of existence and what are its causes ? What means do they propose for transcending existence and achieving liberation from this evil? We explore these issues and conclude the chapter with a critical commentary on these and related aspects of the religion in question.

Hinduism and Moksha

Hinduism has a history of nearly 4000 years during which it has undergone many changes and developments. It is also a vast and unwieldy religious complex with a variety of beliefs, practices and deities. There are, however, some basic concepts notably karma, dharma, samsara or the transmigration of souls and moksha, and beliefs associated with them which constitute Hinduism's identity and continuity over time (Zaehner 1966; Herman 1991). The Varna or the caste system constitutes the social organization of Hinduism. We should note that unlike major religions such as Christianity, Islam and Buddhism, Hinduism has no individual founder to whom the basic teachings could be attributed. It is primarily the sacred texts and the commentaries on them that constitute the chief sources of knowledge for us. Our interest here is essentially in what is known as classical Hinduism (sometimes referred to as Brahmanism to distinguish it from the body of Hinduism as a mass religion), based on the teachings of the Upanishads, which espoused a form of pantheistic monism and whose primary goal was moksha or liberation. Our chief sources of knowledge here are the Vedas and the Upanishads – spanning roughly 1500 to 500 B.C. To these we might add the somewhat later work, the Bhagwadgita (Gita) or the Song/Teachings of the Lord, which forms a part of the epic story of Mahabharata. These works and commentaries on them comprise the basic philosophy and teachings of Hinduism. The earliest works are the Vedas followed by the Upanishads, which span roughly 800-200 BC whilst the Gita is a later composition dating around 200 BC.

The early Vedic age was characterized by optimism and the

affirmation of life. But by the time of the Upanishads it gave way to a preoccupation with life as suffering and bondage (Koller 1982, 67). Thereafter the Upanishads became more and more extreme in their 'revulsion against never-ending life through never-ending death in a manifestly imperfect world ' (Zaehner 1966, 61). Thus according to one of the later Upanishads the human body 'is fair in appearance only; in truth it is no more than a conglomerate of foul-smelling impurities'. And as to the soul it is 'fouler still'. In it are "desire, anger, covetousness, delusion, fear, depression, envy" and human beings are subject to "hunger, thirst, old age, death, disease, and sorrow" (61). It was , above all, the senseless and endless prolongation of life through ever recurring births and deaths - due to the transmigration of souls - that was considered as the evil from which liberation was sought' (63).

In Hindu philosophy, the goal of liberation from existence was closely connected with a number of key ideas, notably rebirth or the transmigration of souls, karma causality and Samsara or the phenomenal world. It was believed that the soul or the self is subjected to repeated births. The 'karma' or the actions performed in each existence conditioned the next birth. Although karma literally means action the reference in this context is to the moral quality of one's actions. Good deeds performed in this life lead to a better birth in the next , e.g. in a higher caste, and conversely evil deeds lead to birth in a lower caste or even as an animal. In any case there was a strict karma causality which causes the soul to return to the earth to 'reap what it had sown' in its previous births. And no matter how good one's actions in life they could not secure him freedom from rebirth. Moksha was deliverance from this sequence of repeated births and deaths or the entrapment of the soul in the phenomenal world or Samsara. Clearly as the quotes above indicate earthly existence, according to the Upanishads, was a pretty sordid business. One life was bad enough but for the soul to be mired in repeated lives was nothing short of a disaster. The world, remarks Zaehner , 'inspired 'a feeling of horror' 'in classical Hindu breasts' (67).

It is important to note here the pantheistic nature of classical Hindu cosmology. Unlike theistic religions such as Christianity or Islam classical Hinduism did not posit a supreme deity as the creator or the master of the world. Rather it was believed that the world had always existed and will always exist. As Max Weber sums up, 'The world,' according to Hindu philosophy 'is an eternal, meaningless "wheel" of recurrent births and deaths steadily rolling on through all eternity' (Weber1968, 167).The only non-temporal orders were the eternal order itself and the beings – the souls – who escape rebirth. Not surprisingly moksha became the central focus of virtually all systems of Hindu philosophy. All of them 'aim at the exposition of the nature of moksha and of the way to realize it' (Lad 1967, Preface).

The essential concept of moksha is the freeing of the soul from its entanglement with the world brought about through karma causality. Although the meaninglessness of existence formed a part of the perception of its evil the literature enumerated other aspects of life which showed its flawed nature. As we shall see later, Buddhism was more precise in its specification of the evils of existence with 'suffering' as its main focus. Hindu thought lacked a comparable emphasis but rather pointed out the 'unwholesomeness, defectiveness and impermanence' of phenomenal existence. Driven by his senses man strove in vain for pleasures, possessions, power and the like which rarely brought lasting satisfaction and in any case it all ended in death. The Upanishads as well as the Gita assume 'the radical defectiveness' of life (Koller 1982, 172). Samsaric existence, according to Gita 'is the embodiment of evil and defectiveness' (ibid.). Lord Krishna, for example, instructs Arjuna that "(earthly) delights are the wombs of dukkha (sorrow)" and that "birth is the place of dukkha"(Herman 1991, 115). Koller (1982, 172) writes of the 'widespread Hindu attitude that pain, fear, anxiety and death are our constant companions' in our 'interminable journey through the cycles of life and death'. Above all what devalued the world in Hindu thought was the *transitory* nature of everything. Earthly life was reduced to naught by the 'metaphysical worthlessness of the transitory, death-consecrated world' with wisdom 'weary of its senseless bustle' (Weber 1968, 170).

Path to Liberation: How could humans be liberated from their samsaric existence ? This became the central question for classical Hindu philosophy. The road to moksha as well as the state of the soul or the self after liberation emerged as a matter of contention among the various philosophical schools. The general answer of the Upanishads seems to be that the main source of human bondage is two-fold: desire or craving for life and ignorance of one's true self. And they are related. Thus the Upanishads taught that 'all things are vanity, and only man is fool enough to desire them. He thirsts for life and the fullness of it, not knowing that it is this very love of life that keeps him a bond slave to the twin evils of *karma* and *samsara* –

samsara that is like a well without water and man the frog that helplessly struggles in it' (Zaehner1966, 63). In other words it was ignorance that made humans a slave of desire and it had to be overcome by knowledge. The human being's instinctual attachment to the world is strong but it needs to be overcome by the higher faculties of reason and self-control which distinguishes humans from the beasts. In the Gita Krishna speaks of "Lust, the ever present enemy of the wise man (jnani)" which "envelops true knowledge (jnana) like an unquenchable fire" (Herman 1991,116).

The knowledge required was that concerning the nature of one's true and deeper self and its relationship to the ultimate reality. According to one of the major schools of thought, Advaita Vedanta, the true self or Atman was identical with Brahman or the ultimate reality. Once this identity was realized the illusory or misleading duality of the self (Atman) and the ultimate reality (Brahman) came to an end and liberation from the samsaric world followed. This "knowledge" was in the nature of an enlightenment or illumination. The quest for this knowledge and self-realization requires inner detachment from the samsaric world and reflection on the deeper levels of one's inner self. That is what leads to the realization of this identity and freedom from desire. The famous Upanishadic proclamation, "thou art that", sums up this self-realization. As Katha Upanishad puts it "A man who is free from desire beholds the majesty of the Self through tranquility of the senses and the mind and becomes free from grief" (Ross 1952, 44). He is no longer bothered by hunger, thirst, sorrow or confusion. He does not worry about old age and death. He experiences "the delight of life and mind and fullness of peace and eternity" (44). Is this symbolism and rhetoric or is it a description of reality? We shall take this point up later. Suffice to say at this stage that Hindu scriptures and philosophies differ in their understanding of the nature of moksha and the state of the soul after it is "liberated" from samsara. The kind of liberation we have described above is the classical Upanishadic view of moksha. It is also the one held by the school of Advaita (monistic) Vedanta. Note that it allows moksha to take place in this life. The one who has conquered desire and realized his true cosmic or transcendental identity is known as Jivanmukta or one who is living and yet freed from life. This concept too will have to be scrutinized later (see below).

Although the path of knowledge or jnana-yoga was considered the superior path to liberation, other paths were also acknowledged later: the path of devotion or Bhakti-yoga and the path of action or karma-yoga. The Bhagwadgita proclaims all three as valid means of liberation. Unlike the pantheism of the Upanishads, Bhakti or devotion is theistic in its conception of the highest reality. In the Gita Lord Krishna reveals himself as a loving and caring God. Wholehearted and unreserved devotion to the Lord can free humans from the grip of samsara. Through the grace of Lord Krishna one's karmic accumulations are nullified and the liberated self enjoys eternal union with the Lord. As Hinduism evolved into a polytheistic religion it extended the same idea to other deities. Karma yoga, the other path to liberation, involves active participation in the world in accordance with one's Dharma or caste duty. But how can this lead to liberation? For is this duty not expected of all Hindus? Yes, but there is a difference. The essential point here is that action has to be *disinterested*, i.e. free of desire. When one acts without desire the action is not binding. There is no karmic accumulation. It is performed in the spirit of inner detachment such that one acts but without any expectation of rewards. It is a form of renunciation *in* action. Through such action or Karma yoga the cause of suffering (dukkha), the craving or desire, is destroyed and one achieves liberation (Herman 1991, 119-21; Zaehner 1966, 102-3).

We turn next to some of the socio-cultural aspects of moksha. Was it meant to be universal, i.e. available to all, or were there differences according to caste and other affiliations? Here we see the importance of caste in Hinduism. In the early Upanishads the attainment of spiritual knowledge and liberation was seen as accessible to Brahmins, the highest caste, alone. The later Upanishads extend the scope to the other two "twice-born" castes, namely the Kshatriyas and Vaishyas. Evidently the lowest caste, namely the Shudras were left out with no hope of liberation from samsara. The Gita, however, appears to universalize its message. Even the Shudras could achieve liberation if they followed one of the three yogas or paths laid down by Krishna. Indeed by extension release from samsara was available to Hindus and non-Hindus alike.

Turning to the social organization of Hinduism we see that apart from the caste system there is also a concept of the principal values or goals of life as well as a division of the life cycle into a number of phases. The four goals of life were artha or material prosperity, dharma or performance of caste duties, kama or seeking pleasure including sexual enjoyment and moksha or spiritual quest and liberation. But how could one seek moksha while pursuing these other mundane earthly goals at the same time? Part of the answer seems to be provided by the life cycle and its stages. The first stage was that of a Brahmacharin, i.e. a youth who is a student, is celibate and is practicing self-discipline. The next stage was that of a Householder, i.e. a married man with children, the third that of a Vanaprasthi, i.e. one who is in the process of disengaging himself from his worldly involvements, and the final one was that of a Sannyasi, i.e. an old man who has freed himself from all social and worldly involvements and is immersed in spiritual thought and practices aimed at moksha. As we can see here the goal of moksha, though acknowledged as the most important of life's goals, has been integrated into the average Hindu (male)'s life[1]. Clearly liberation has to be sought after the individual has done his worldly duties including the production of offspring, especially sons, required to perform certain religious rites following his death.

While it has to be recognized that the four goals of life as well as the four stages in the life cycle are a schematic and idealized versions of Hindu way of life their material and social concerns are also clear. Here one may speak of a contradiction between the radical rejection of phenomenal life as virtually worthless, if not downright evil, on the one hand (as we saw above) and on the other the practical down-to-earth organization and reproduction of life as part of a stable social order as indicated by these schematics. To some extent the gap between these two "worlds", as it were, can be bridged by the notion of a mass and a virtuoso religiosity (Weber 1963, 174). It seems that for the masses moksha remains a "part-time" or end of life pursuit since the business of life must go on. On the other hand there are plenty of indications in the literature to suggest that a minority may seek enlightenment and liberation without becoming Householders (174). While there is a general presumption of marriage and children as a part of normal life, celibacy, asceticism and the renunciation of worldly existence – chiefly a domain of the virtuoso - has been valued highly in Hindu society and culture. Another way of looking at this "contradiction" is to distinguish between Classical Hinduism as a "doctrine of salvation" (Upanishads) and Hinduism in general as a religion serving the social and spiritual needs of people. As Vallee-Poussin (1917, 3) observes, unlike religions these disciplines of salvation are 'made for ascetics only...they are purely personal or individualistic...unsocial and often antisocial: they deprecate and often prohibit marriage' . Thus we read in Brihadaryanaka Upanishad "Wishing for that world (for Brahman) only, mendicants leave their homes....Knowing this, the people of old did not wish for offspring. What shall we do with offspring, they said, we who have this Self and this world (of Brahman)..... And they, having risen above the desire for sons, wealth, and new worlds, wander about as mendicants" (Max Muller 1884, 179-80).

The Nature of Moksha: In any case the goal of liberation can be pursued, albeit in different ways, by both the householder and the world-renouncing monk. And at least according to some of the Indian schools of philosophy, notably Advaita Vedanta, liberation can be attained during one's lifetime. But how does one know when one has attained moksha ? An answer to this is that one "knows" this intuitively when one reaches the highest levels of consciousness. The realization, the deeply felt experience of the oneness of the Atman with the immanent Brahman authenticates this "knowledge" (Ross 1952, 21). Presumably the Jivanmukta, after attaining this state still goes about his life in the normal way but being free from desires he sees the world differently. Put simply he is "in the world but not of it". In the eloquent prose of Hindu scriptures "as soon as the individual self has acquired the perfect immediate certainty that he is the universal Self, he no longer experiences doubt, desire or suffering. He still acts, as the wheel of the potter continues to revolve when the potter has ceased to turn it. Death at last, abolishes what no longer exists for him, the last appearance of duality" (Vallee-Poussin 1917, 28). Elsewhere it is asserted, "When desire ceases, the mortal becomes immortal; he attains Brahman on earth" (140). According to Advaita Vedanta, moksha is a state of pure being, pure consciousness and pure bliss, i.e. Satchitanada (Zaehner 1966, 76). However accounts of the state of moksha vary and one answer is that it is beyond the realm of thought and expression. With the subject-object distinction obliterated "it cannot be designated." "It causes the phenomenal world to cease"(75). In the words of Gita the liberated man "Seeing himself in all things and all things in himself, he sees the same thing everywhere" (94). He is beyond pleasure and pain and beyond the sense of I' and 'mine' and all the opposites (94). Knowing that his true being is outside space and time he conquers death (94). It is also assumed that with the realization of one's true self one is also freed from the cycle of rebirth and redeath. One's 'accumulated karma is destroyed' and one 'does not acquire any fresh karma'.

What happens to the soul or the essential self after moksha? Where does it go? Does individual identity cease or does it continue in some form? The answers to these questions are somewhat diverse. The position of the pantheist philosophies, adhered to largely by the Upanishads and such leading schools of thought as Advaita Vedanta and Samkhya, is that liberation ends the empirical or phenomenal self. According to Advaita the illusory duality of Atman & Brahman ceases with the realization that they are identical. The result is that the immortal soul returns to its pure being in a state of full consciousness and bliss. With Samkhya, the soul, freed from the body, returns to its own eternal self-reflective consciousness (Encyclopedia of Hinduism, 379). For the Nyaya-Vaisesika school the soul returns to a state of unconscious and indifferent pure existence, 'like a stone' (Lad 1967, 6). Unlike these pantheistic doctrines the theist philosophies, including the one elaborated in the Gita by Lord Krishna himself, teach that the liberated self-survives. It is immortal and enjoys eternal bliss in the heavenly world of satisfied desires and undreamed of delights in the presence of Lord Krishna (Herman 1991, 118). In fact the Gita is somewhat ambiguous in that it also expounds the Upanishadic view of moksha, i.e. the absorption of the Atman into Brahman (118). In this case the devotee's ego, personality, mind, memory and consciousness are all assimilated. According to the later polytheistic views Brahman can be personalized as god. After his death the devotee attains the realm of the god he worshipped and enjoys blissful communion with his deity (118-9). Clearly with the introduction of Bhakti yoga or devotion to a personal god as one of the paths to liberation it is not surprising that the devotee is rewarded with eternal presence in the proximity of his adored deity.

There seems to be a clear divide between the pantheistic and the theistic doctrines with respect to the post-moksha state of being. The former see the disappearance of the empirical self and the return of the immortal soul to its ground of being, free from all extraneous relations. The latter see the continuation of the self in some form or the other in the graceful presence of god. In all cases, however, one thing is certain. He who attains moksha is liberated from samsaric existence. He has secured his release from the recurring cycle of birth, decay, suffering and death.

<u>To Summarize</u>: According to classical Hindu beliefs, as articulated in the Upanishads and the Gita, existence in general and human existence in particular is fundamentally flawed. From the beginning to end, i.e. from birth to death, life involves suffering. Moreover everything in the phenomenal world is transitory - ephemeral and passing. Human consciousness or the soul finds itself 'trapped' in material existence and dragged through an endless cycle of birth, decay, old age and death with apparently no rhyme or reason and with no end to the process in sight. Moreover earthly existence was found wanting on both moral and metaphysical grounds. Immorality was rooted in the very instincts and desires that drive man's existence while the impermanence and transitoriness of the world of phenomena deprived it of any metaphysical worth. Indeed at least one school of thought, viz. Advaita Vedanta, found the phenomenal world so outrageous as to suggest that it was unreal and illusory. The Maya or illusion involved was due to ignorance of true reality and Gnostic knowledge was the necessary cure.

The radical devaluation of the phenomenal world and its rejection was coupled with the quest for transcending existence and anchoring the essential self or the soul to a permanent and imperishable reality free of temporality and of all material exigencies and suffering. Belief in the existence of a 'soul' that is immortal and partakes of the absolute but inhabits the body led to ways and means of realizing the true nature of the immortal self and freeing it from its earthly bondage. This was the spiritual quest, which could be pursued in a variety of ways, leading to moksha or liberation. Clearly the 'knowledge' or experience of moksha was ultimately a mystical experience which the liberated individual could not communicate to others. And as for release from transmigration the knowledge that the liberated soul will not be reborn was a matter of faith for which there could be no external 'proof'. Clearly the radical rejection of the world and the single-minded quest for liberation was meant for the few – the intellectuals and the religious virtuosi, those who found the world wanting – rather than the many.

For the masses the continuation of worldly existence through marriage and reproduction was enjoined as the normal way of life. However for the Householder too moksha remained the ultimate goal but it was to be pursued at the last stage of the life cycle after the individual had completed the 'normal' business of living. We should note that the concept of moksha or liberation had two aspects – one 'negative' and one 'positive'. There was a great deal of unanimity about the former, i.e. what one was being liberated *from* but far less about the latter, i.e. what one was liberated *to*. Among pantheists, at least one school of thought equated liberation virtually with oblivion while others envisioned a peaceful, tranquil or joyful state of timeless consciousness for the liberated soul. Theists believed in a blissful state of union or communion with a personal god in a heavenly abode. As religious belief systems, all of them included a 'spiritual' dimension, the experience of a deeper or higher level of reality. The rejection of the phenomenal world of mundane existence and liberation from it did not mean an acceptance of nothingness. Rather a positive spin was put on moksha. Whether in this life or post-mortem, the apprehension of Brahman or the cosmic ground of being, the experience of a mystical state of consciousness, the ecstatic devotion to a personal god, an immortal heavenly existence – all these represented a higher level of 'living'. Repudiation of the ordinary worldly existence was not to be equated with embracing nothingness.

It appears that the sacrifice and the hardships involved in the quest for liberation was to have its 'reward'. For a religious doctrine it was difficult, if not impossible, to posit total annihilation or disappearance of the 'self' or the soul. As we shall see below even Buddhism, a religion which comes close to being atheistic also held out the promise of nirvana being a blissful state. Schopenhauer, who believed that his concept of the renunciation of the will resembled the Hindu and Buddhist notion of rejecting existence, cautioned that "We must not evade it, as the Indians do, by myths and meaningless words" but openly acknowledge that the end result is simply nothingness (Nicholls 1999, 175).

Buddhism and Nirvana

Buddhism grew out of the same Indian metaphysical soil as Classical Hinduism. The period in which Buddha, the former Prince

Siddhartha, preached his doctrine of liberation, namely the 6th century BCE, was a period of great philosophical and religious ferment. Out of this ferment emerged two new religions, viz. Jainism & Buddhism. Although Buddhism died out in India it flourished abroad becoming one of the world religions. Schisms and various schools within Buddhism developed later, the main divide being between the Mahayana and the Hinayana or Theravada branches. The former opted for a mass religiosity which emphasized compassion and salvation for all rather than an ascetic, monk-centered religion of withdrawal seeking nirvana (Koller 1982, 164-7; Snelling 1991, Chs. 7, 9). However our concern here is with the early or Classical Buddhism, the one before the schism. As with Hinduism, we are mainly interested in Buddhist attitude to existence, its specification of the problematic of existence, as well as the nature of and the road to liberation.

Compared with the teachings of Classical Hinduism, expounded in the Upanishads and the Gita, Buddhism represents a more logical, coherent and systematic analysis of the problematic nature of existence and its solution. However it shares with Hinduism some key assumptions, notably the belief in samsara or the interminable cycle of rebirths, and karma causality. Moreover nirvana, its concept of liberation, is similar to the Hindu concept of moksha, at least in its key defining characteristic, viz. freedom from rebirth.

Buddha's first sermon lays out the basic principles and the framework of his teaching in the form of *Four Noble Truths*: that suffering exists, that it has an identifiable cause, that this cause can be removed, and that it can be done by following the Eight-fold path.

"Suffering I teach and the way out of suffering" was Buddha's ringing declaration, the *First Noble Truth*. Clearly the fundamental feature of existence, for Buddha, is Dukkha or suffering. It is all-pervasive and ever-present. Dukkha has a wider connotation than suffering and includes not only pain but also sorrow, unhappiness, disappointment, regrets, worry, unease, a sense of malaise and similar negative states. Each living being seems to be in bondage to the sources of life's suffering. How to break this bondage? This is the central problem the Buddha addresses and this leads on to the *Second Noble Truth*, i.e. the underlying cause of suffering . Buddha identifies it as Tanha (trishna in Sanskrit) or desire. Trishna literally means thirst, in short craving for worldly things and pleasures. This thirst seems natural to us. It is implanted in us as the will-to-live .But since it is the source of our suffering, and its endless prolongation through repeated births and deaths, it is this that we need to liberate ourselves from. But how? This leads to the T*hird Noble Truth*. The cessation of suffering requires that we free ourselves completely from the bondage to desire and craving. And this is what the *Fourth Noble truth* is about. It sets out an eight-fold path leading to the conquest of desire and liberation from suffering and death. The path emphasizes right knowledge or understanding of the human condition, moral conduct and self-control, and meditation as the three principal means of overcoming life's bondage. Each individual has to seek and win his own liberation in accordance with the Buddha's teachings. The ultimate goal is nirvana or the extinction of the flame of desire, a state of being which ensures that the liberated one, the Arahat, will no longer be subject to rebirth (Koller 1982, Ch. 7; Snelling 1991, Ch.7).

<u>The four Noble Truths:</u> These truths present us with an outline of the Buddhist perspective on existence and the pathway to liberation. However we need to look at all four in some detail in order to grasp their meaning and implications. *First* and foremost, existence is seen by the Buddha essentially as suffering. Here, perhaps not surprisingly, The Buddha's perception of life is echoed, in faraway Greece, by Socrates who is credited with the remark "to live is to be sick for a long time". And this medical metaphor is in fact also used by the Buddha. Suffering is likened to a disease for which the Buddha offers a diagnosis and a cure. Here is how the Buddha elaborates on the theme of suffering. "This, monks, is the Noble Truth of suffering, birth is suffering; decay is suffering ; illness is suffering; presence of objects we hate is suffering; separation from objects we love is suffering; not to obtain what we crave is suffering; In short, the five attachment groups are suffering" (we explain the meaning of the attachment groups later). And here is how he castigates birth, which ushers in our lifelong suffering: "Shame on this thing called birth", for it brings in its train "decrepitude, disease and death"(Koller 1982, 141-2)). Clearly the evil of existence begins with birth (Dahiya 2008, 98).

But is the Buddhist perspective not entirely 'one-sided', in short a pessimistic view of life?' How could the Sakyamuni arrive at this conclusion? As a former prince who led a sheltered and charmed life for the first 29 years of his life he must surely have known life's many comforts and pleasures? He was married and had a young son. So he knew that alongside Dukkha or suffering there is also Sukha, or joy and happiness, in life. Indeed, he did recognize this side of life but

32

considered these felicities as few and far between and, in any case, they were short-lived and transitory. They did not last. In fact for the Buddha the impermanence and the transitoriness of everything - a feature of phenomenal existence – is also an aspect of suffering. We want to hold on to what is dear to us yet change and dissolution is in the nature of things. The emphasis on suffering means that for the Buddha it constitutes the hallmark of existence (Snelling 1991, 52-3). Overall it is Dukkha or pain and suffering that is the central fact about life.

Elsewhere Buddhist doctrine points out three distinct marks of existence, viz. Dukkha or suffering as stated in the first Noble Truth; impermanence or the transitory and changeable nature of everything; and lastly the absence of a 'self' or essence in everything that exists. Thus in addition to the directly experienced pain and suffering, the impermanence or the unreliable and ephemeral nature of phenomena, and the absence of a 'self' or essence in everything devalue existence further (Snelling 1991, 64;Koller 1982, 146-8).

, Let us go back to the five attachment groups mentioned in the first noble truth which are related to this lack of an essential self. The doctrine of attachment groups is meant to expose the delusion of an 'I' or 'mine', an 'ego', the sense of possession and individuality which is crucial for sustaining our craving and desire. The human being can be seen as comprising five groups of elements: the physical body, sensations and feelings, cognition, character traits and dispositions, and consciousness. These together constitute the 'person' or self. But the point is that each of these five factors is constantly changing. They are variable and ultimately perishable so that they cannot provide a solid basis on which to build a secure and satisfactory life. The statement that the five attachment groups are suffering refers to this fundamental reason why human life can never be ultimately satisfying. For apart from these five groups of traits and their interaction there is no 'self'! Once we realize that over and above these variables there is 'nothing' else then we can be free of our sense of ego and the striving for worldly things. It is our *attachment* or clinging to these changing and impermanent entities that is both a source and a form of suffering.

We turn next to the *second Noble Truth* which is about desire or craving, the cause of our suffering. Here is what the Buddha has to say: "This, monks, is the Noble Truth concerning the origin of suffering: it originates in that grasping which causes the renewal of becomings (rebirths), is accompanied by sensual delights, seeking satisfaction now here, now there; the grasping for pleasures, the grasping for becoming (existence), the grasping for non-becoming (non-existence)" (Koller 1982, 135). The 'non-becoming' presumably refers to the desire for the non-existence of things, people and conditions and may also include the urge to suicide. Whatever form it takes it is this trishna or desire, this craving 'that causes rebirth' (Keown 1996, 52). It is about the instinctive will-to-live which drives us on to new lives and new experiences. Clearly this formulation strikes at the very root of our existence in so far as it traces back the cause of suffering to the innate will-to-live and the striving after the myriads of things that life has to offer. It is this thirst that needs to be quenched once and for all.

Some modern day interpreters of Buddhism tend to play down the wholesale condemnation of desire or the will to live in Buddhism. Keown (52) for example writes that the Pali word Tanha (Trishna in Sanskrit), translated as thirst or desire 'connotes desire that has become perverted in some sense, usually by being excessive or wrongly directed'. He argues that we can distinguish between good or right desires and bad or wrong ones (52) and it is the latter that the Buddha is concerned with. The point is that Buddhism as a religion has moved a long way from the austere teachings of its founder, and in any case there are numerous interpretations of the Buddha's teachings. Nonetheless it seems that Keown's distinction between right and wrong desires is difficult to sustain, at least from what is known of the Buddha's own sermons and teachings. As well, most interpreters of tanha seem not to make any such distinction. Ancient Buddhism, it appears refers to the *normal*, in-built life force emanating from innate drives. Thus Herman writes: 'The cause of suffering is existentially grounded in the individual' (Herman 1983, 59). Or as Snelling puts it: 'Basically tanha can be reduced to a fundamental ache that is implanted in everything that exists, a gnawing dissatisfaction with what is and a concomitant reaching out for something else. So we can never be at rest but are always grasping for something outside ourselves. This is what powers the endless ....Wheel of Life' (Snelling 1991, 53). In his Fire Sermon the Buddha spoke of all human life as 'ablaze' with desire. Fire is an apt metaphor for tanha or desire, since it grows on what it feeds without ever being satisfied (Keown 1996, 51). To escape suffering we have to break free of desire. Thus all grasping, all passions must be "laid aside, given up, harbored no longer and gotten free from," proclaims the third Noble Truth (Keller 1982, 3). How to achieve this difficult objective is the subject matter of the fourth Noble Truth which lays out an eight-fold path to achieve liberation from desire and attain nirvana. The Path consists of (1) Right View (2) Right Resolve (3) Right Speech (4) Right Action (5) Right Livelihood (6) Right Effort (7) Right Mindfulness and (8) Right Meditation.

A Doctrine of Salvation: The pattern of thought and action indicated by the Path is, in principle, open to anyone. But in reality its proper and full realization seems to require a withdrawal from worldly activities and a total dedication to achieving liberation or nirvana. It is generally acknowledged that Buddha's teachings represent essentially a 'doctrine of salvation' meant for the world-renouncing monks rather than lay people. Although Buddha's followers included lay people, soon a clear distinction was drawn between the laity who were not seen as qualified to achieve nirvana and the monks whose way of life alone was thought to provide the conditions necessary for liberation. These included non-possession of money, living on alms, homeless or mendicant status as distinct from the laity who were 'house-dwellers,' and strict celibacy. As Vallee-Poussin (1917, 150) observes, 'The only Buddhist, in the proper meaning of the word, is the monk who has broken all ties with society'. The same point is made by other commentators. Buddhism was meant essentially for the monks, 'for those who had retired from the world of activities to lead a celibate and monastic life' (Herman 1983, 64). Or as Max Weber (1967, 214) the great sociologist of religion wrote, 'Wandering homelessly, without possessions and work, absolutely abstemious as regards sex, alcohol, song and dance..... living from door to door by silent mendicancy, for the rest given to contemplation', such was the way of the Buddhist seeking ' salvation from the thirst for existence'[2]. As for the laity, certain moral precepts were stipulated whose observance was to be rewarded by material well-being in this life and a better condition of life in the next birth. As a primary generator of passion, sexuality was considered 'extremely dangerous by the Buddha' (Snelling 1991, 58-9). The laity were to exercise moderation in sex as 'indulgence of sexual desire could only serve 'to feed the fires of passion and attachment' (59). An important duty of the laity was to provide material support for the monks. It was indeed the highest honor and merit available to the lay or "house-dwelling' Buddhist (Weber 1967, 215, 219).

Clearly the radical devaluation of the "world" and withdrawal from it into a mystical state of contemplative stance through meditation, implicit in the eight-fold Path, was something that only a small number of dedicated individuals – the virtuosi, the salvation-seekers – could hope to achieve. Buddhism began almost exclusively as a doctrine of salvation and gradually developed into a mass religion. The schism following Buddha's death between Hinayana or Theravada and the Mahayana schools reflected this development. The latter, among other things, moved towards a religion which served the needs of the lay people, emphasizing faith and compassion. The former remained closer to the stance of ancient Buddhism, viz. a religion essentially for the monkhood.

What is Nirvana? The ultimate goal of Buddhist quest is nirvana which can be achieved by following the eight-fold path. But what exactly is nirvana and how does one know when one has reached this blessed state? Seemingly these are simple questions that have, to put it mildly, no clear answers. The mystique and the ambiguity surrounding the concept of nirvana are reflected in the vast literature that has grown up around the subject. Perhaps more has been written on nirvana than on any other religious concept. Buddha himself said little about nirvana, at least directly, and discouraged speculation and theorizing about it, e.g. what was the nature of nirvana-in-life, nirvana after death and so forth. When pressed for an answer concerning these and similar questions Buddha's reply was "Whether this or that dogma is true, there still remain birth, old age, death, for the extinction of which I am giving instructions…What I have left unsettled, let that remain unsettled" (Vallee-Poussin 1917, 130).

Nirvana literally means 'extinction' or 'blowing out' as of a flame, and this seems to be an appropriate metaphor in this context. It denotes the extinction of the flame of desire and craving, the thirst for life which creates attachment and rebirth. The liberated one, the Arahat, is one who has finally transcended all craving and desire and who has direct intuitive knowledge of having done so. Such a one may be said to have attained nirvana in life. Freed from all karmic consequences he will not suffer rebirth. At last he or she has been liberated from the incessantly turning wheel of life.

The fire of which Nirvana is the extinction is described in Buddha's 'Fire Sermon'. It pertains to the three inner fires of greed, hatred and delusion and the three external fires of birth, aging and death. Nirvana during life is frequently described as the destruction of the three ' fires' or defilements. One who has destroyed these cannot be reborn and so is totally beyond the remaining 'fires' of birth, ageing and death, having attained final nirvana (Harvey 1990, 61). The state of the Arahat who has achieved deliverance is said to be one of great inner peace, tranquility and contentment, in a certain sense a state of bliss. Nonetheless speculation and controversy goes on about the state of nirvana in this life and beyond. As in the case of moksha one approach is to suggest that it is essentially 'mystical' in nature and so cannot be conveyed in words. Thus descriptions of nirvanic experience stress its "otherness," 'placing it beyond all limited concepts and ordinary categories of thought' (62). In the face of nirvana 'words falter, for language is a product of human needs in this world, and has few resources with which to deal with that which transcends all worlds' (62). What happens to the one who is freed from rebirth *after* death is also a matter of speculation. For unlike Hinduism, Buddhism rejects the idea of a soul or self which survives death. Yet it does believe in a process of karma causality which begs the question of who or what is the vehicle of karmic consequences and which undergoes rebirth and transmigration. This was another question the Buddha refused to answer holding it as irrelevant to the problem of Dukkha and deliverance from it which was the substance of his teaching.

To summarize: The Buddhist view of existence is very similar to that of Hinduism. Both see it at the very least as an undesirable, if not an evil, state and look for a way out. The Buddha articulated this viewpoint quite clearly and forcefully, defining existence as suffering, and providing a systematic and detailed analysis of the cause and cure of this malady. Common to both religions is the belief in the reincarnation of beings in an endless process of rebirth and redeath.

And it is this process of samsara, in particular, from which release is sought. To get off the perennial treadmill of birth, decay and death is the supreme goal of both, expressed as moksha in the one case and nirvana in the other.

Scholars and others reflecting on the "pessimism" of these doctrines have wondered if it was not so much life per se but the timeless cosmic process of samsara, i.e. dying and being reborn repeatedly, the "eternal recurrence," that drove these two belief systems into seeking an escape from existence. However this does not seem to be the case. For if existence were considered to be a "good" then recurrence of births and deaths should have been welcomed since rebirth means a new lease of life and death is not the end because it is merely a prelude to a new birth even if in a new life form. In short, reincarnation can be seen as a form of immortality. Now if these doctrines approved of worldly existence then why should they seek an exit from it? Surely it would be in their interest to perpetuate rebirth and their main concern should be to preach those moral precepts whose observance ensures good karma and helps the faithful achieve a good rebirth. Indeed, this has been a part of the teaching aimed at the laity by both religions. However, their *summum bonum* is not higher rebirth but liberation from the samsaric process altogether.

Another relevant issue is that both Hinduism (at any rate some of the major schools) and Buddhism also see the possibility of liberation–in-life although the liberated individual cannot communicate this experience to others in any way since it lies beyond all categories of worldly existence. In fact there is a great deal of ambiguity about what it means to be 'liberated- but-living.' The best accounts see it as a mystical state of blissfulness. Now if it is a condition of blissfulness, an extraordinary state of being that can be experienced in this life, then arguably it would be desirable for the individual to be reborn to have an opportunity to experience such a state again. But clearly that is not how these doctrines see the situation. Liberation-in-life is seen as only a stage in the ultimate liberation from rebirth. Thus existence in all forms, including in the blissful state of moksha or nirvana while living, is rejected in favor of freedom from rebirth altogether. This brings us to the inescapable conclusion that both these perspectives see worldly existence as an undesirable state, shot through with negativities of all kinds, so that escaping it altogether is the highest good possible for human beings. Even the blissful state of moksha or nirvana, which can be experienced while living, through individual effort, is not enough to justify rebirth.

Finally, for both religions transcending existence involves mysticism and faith. Thus liberation-in-life means turning away from worldly desires and preoccupations and escaping into higher reaches of being through immersion into a spiritual realm. This is a form of mysticism. The other form of liberation from existence is post-mortem, viz. the promise of freedom from rebirth. This is a matter of faith, i.e. that there is such a thing as rebirth and that one would escape it.

Endnote – Chapter 1

1. Was moksha meant for men only? This appears to be the case given the subordinate status - that of a server - accorded to women (Koller 1982, 73-6). Moreover the assumption seems to be that only *men* – and a small minority at

that – are capable of attaining the knowledge, the consciousness, the discipline and the asceticism required to free them from worldly attachment. Given this general stance of Hindu scriptures it would be misleading to replace 'he' with 'she' in the text for the sake of politically correct language. However with the introduction in the Gita of Bhakti yoga or devotion to Lord Krishna - and by extension to other deities - as a path to liberation the concept of salvation became more inclusive and presumably applied to female devotees as well.

2.   As would be evident from the stringent requirements for Buddhist monkhood, including the state of an itinerant individual, it was ill-suited for women. However the Buddha's essential message was egalitarian in its general orientation, e.g. in its rejection of caste distinctions, and was meant for all. Women were considered by the Buddha as capable of achieving the state of an Arahat and attain nirvana. Thus in ancient Buddhism, in principle at any rate, nirvana could be achieved by anyone prepared to follow the eight-fold path. The Buddha accepted the ordination of women as nuns and endorsed the formation of the order of nuns similar to those of monks although the nun's status remained inferior to the monk's. Subsequent development of Buddhism, with its schism and various socio-cultural influences, changed the situation in many ways. For example the order of nuns virtually disappeared in Theravada Buddhism only to be revived recently.

Chapter 2

Philosophical Perspectives: 19th Century

Secular philosophies of existence are primarily a Western phenomenon. But concern with existential issues has not been the hallmark of Western philosophy. Almost since the time of Plato but particularly since Descartes, who is considered as inaugurating the modern age of philosophy, it has been the preoccupation with knowledge. Questions such as what do we know about external reality? How can such knowledge be authenticated? What is the role of sense perception in all this? And how does human mind or reason relate to knowledge? In short epistemology has been the chief focus and concern of Western philosophy for more than three centuries. Other issues with which it became involved in the 20th century were about language and 'meaning'. This is especially true of philosophy in the English-speaking world. Thus 'to what do we know', was added the question 'what do we mean', e.g. when we say so and so? Meanwhile problems of human existence, e.g. the fundamental characteristics of existence itself, the meaning of existence for human beings, and the wider implications of all this, remained peripheral if they were not ignored altogether (Magee 1978, 77-81; Benatar 2004, 1-2). It was not until the 19th century that we see the beginning of engagement with existential issues. In different ways the works of

Schopenhauer, Kierkegaard and Nietzsche represent this new departure. Of these it is only Schopenhauer, one of the great philosophers of all time, who takes a rejectionist view of existence.

In the 20th century existential issues were taken up by other European (Continental) thinkers, notably Heidegger and Sartre, who reflected systematically on the nature of man's being in the world. After WWII the catch-all phrase 'existentialism' came to symbolize this new current of thought, mainly centered on the European Continent. We shall examine rejectionist philosophies of the 20th century and beyond in the next chapter. For the moment our focus is on rejectionist thought in the 19th century. Undoubtedly Schopenhauer is the key figure whose 'pessimistic' philosophy was highly influential especially in the Germanic world. Eduard von Hartmann was one of the thinkers who came after Schopenhauer and was undoubtedly influenced by him. His philosophy has many similarities as well as some significant differences with that of Schopenhauer. Hartmann was a highly regarded and popular thinker of the last quarter of the 19th century. In this chapter we shall look at these two philosophers.

Arthur Schopenhauer: Suffering and Willlessness

It is interesting to note that Schopenhauer was the first Western philosopher who was a professed atheist (Magee 1988, 213). Almost all his predecessors, with the notable exception of David Hume, were theists of one kind or another. They all posit a God or Godlike presence behind the phenomenal world. Human existence, with man as the unique possessor of reason, was glorified or just taken for granted and rarely, if at all, questioned as problematic in any fundamental way. A well-known example of this glorification or at least justification is Leibnitz's view that God has created "the best of all possible worlds". It was Schopenhauer who radically broke with this trend and propounded a metaphysical doctrine which saw existence largely as a source of pain and suffering for all creatures. As we shall see, this is related to his concept of the Will - will-to-life in the case of living things - the driving force behind all existence. Will-to-life entails a struggle for existence within and between the species as well as the reproduction of the species. The result is the perpetuation of pointless misery and suffering; pointless because there is no ultimate aim or goal beyond the maintenance of one's own existence and the perpetuation of life through reproduction. The pointlessness of the blind Will, with its constant activity without any rhyme or reason, is more clearly evident at the cosmic level. Schopenhauer's relentless and uncompromising view of existence as nothing short of evil – something that should not be – has earned him the reputation of being the 'pessimist' philosopher par excellence. His philosophy has striking parallels to Buddhism albeit it was arrived at without any prior knowledge of the latter. We begin this section with Schopenhauer's view of existence and what he sees as the path to liberation.

Will and Suffering: The fundamental principle underlying all existence, according to Schopenhauer, is the will-to-life or simply Will. This is not the conscious willing that we think of when we use the word 'will'. Rather it is a blind, innate urge or force that drives all existing things, animate and inanimate, towards some end or the other. For living things, it means above all the urge to survive and to reproduce and perpetuate the species. This is the basic nature of existence that humans share with all living things and, as Schopenhauer sees it, it involves untold suffering. The similarity with the Buddhist view of existence is striking. The Buddha's First Noble Truth is that life is suffering. Schopenhauer arrived at his perspective on existence without any knowledge of Buddhism or Brahmanism. Later, when he discovered the Upanishads and the Buddhist literature, he was struck by the common ground between his reflections in this regard and Indian thought, particularly Buddhism (Schopenhauer 1977, v.i, xiii; v. ii, 371) Schopenhauer was gratified to see his viewpoint corroborated by these two ancient religions, which he believed had confronted the reality of existence without illusions and made liberation from existence as their supreme goal (Schopenhauer 1969, V.ii, 627-9).

For Schopenhauer pain and suffering are intrinsic to life in a fundamental way. To begin with life literally feeds on life. We see this clearly in the animal world. Animals have to devour other animals, plants and organisms in order to survive. The pain and suffering of life is revealed poignantly in the death agony of the helpless prey struggling in the jaws of its predator. We humans also devour life – plants, fish, birds and other animals – but are mostly unaware of or turn a blind eye to this essential foundation of our existence. We rationalize this act of cannibalism by the pretence that our killing of animals for our nourishment is done humanely or that it is justified since they are a different species from us. This, for Schopenhauer, is mere self-deception or a lack of awareness concerning the immorality and cruelty underlying our existence. It simply means that humans have established their domination over the rest of nature and think that all of nature is simply there to serve their wants and desires (Schopenhauer 1969 v.i, 146-7).

In drawing our attention to the sufferings of the animal world, inflicted by animals on each other and by humans more systematically, Schopenhauer breaks new ground. Neither Brahmanism nor Buddhism paid any attention to animals but focused entirely on human suffering. Western tradition of thought in general has been even more clearly anthropocentric. As Schopenhauer points out, Christianity assigns humans a privileged position in God's creation, making them the lord and master as well as the custodian of the world. Indeed so eminent a moral authority as Immanuel Kant considered it appropriate to treat animals as means to our ends. Since they were devoid of reason they were merely 'things'. Hence Kant declares that 'man can have no duty to any beings except human' (Murdoch 1993, 253). Schopenhauer traces the source of Kant's attitude to Christianity, since 'Christian morality leaves animals out of account'. Even the great Christian mystic Meister Eckhart writes to the effect that "all creatures are made for the sake of man" and that all "created things become of use to the good man" (254). True, at least one of the ancient Indian religions, namely Jainism, had made non-violence ('ahimsa') to all living creatures one of their cardinal moral precepts and adopted strict vegetarianism. Nonetheless Schopenhauer's philosophical viewpoint, which looks upon the suffering of all living beings and not just those of humans, seems to be unique in drawing our attention to the fact that all living creatures including human beings are by the very nature of their existence involved in killing others or getting killed. Not surprisingly Nietzsche described Schopenhauer as "the only serious moralist of our century" (57).

Apart from the killing of living beings to sustain life, there is also a constant struggle for existence between species and between individuals within species. Conflict over possession and domination is endemic in life. Inflicting pain and suffering on each other is thus intrinsic to living things. The history of humanity is a saga of conflict and struggle – between individuals and groups such as tribes, nations and classes – as reflected in our daily lives and in hundreds of wars, rebellions and revolutions throughout history. Schopenhauer alludes to the unimaginable barbarism and cruelty to which African slaves were subjected to for several hundred years (Schopenhauer 1970, 138). The suffering inflicted on the native peoples of the Americas by the conquistadores is another of the innumerable examples of the unspeakable cruelty and suffering inflicted on human beings by their fellow-humans. To this man-made evil, stemming from the nature of the will-to-live, we have to add the death and destruction resulting from natural disasters, e.g. floods, cyclones and earthquakes. As Schopenhauer remarks, 'If the immediate and direct purpose of our life is not suffering then our existence is the most ill-adapted to its purpose in the world' (41). And although Schopenhauer's focus is always on the will-to-life and its consequences, natural disasters such as volcanic eruptions and earthquakes also feature as a source of suffering stemming as they do from the Will which animates natural forces. Perhaps this more comprehensive vision is what he has in mind when he writes, 'If you imagine….the sum total of distress, pain and suffering of *every kind* (italics added) which the sun shines upon in its course, you will have to admit it would have been much better if the sun had been able to call upon the phenomenon of life as little on the earth as on the moon; and if here as there, the surface were still in a crystalline condition' (47).

Preponderance of Pain over Pleasure: Schopenhauer's emphasis on pain and suffering of existence raises the question of his neglect of the 'other' side of life, viz. the pleasures, joys and satisfactions that are also a part of existence. Is there not, then, a trade-off between pain and pleasure, the joys and sorrows of life? This brings us to a complex series of arguments that Schopenhauer advances in defense of his viewpoint. First of all, he maintains that the pain and suffering involved in human existence far exceeds its pleasures. This is, in part, because we take normal well-being and satisfaction for granted and do not feel particularly joyful about it. On the other hand pain and suffering register upon us far more strongly and immediately (41-2). For example the fact that I am free of toothache at the moment does not make me feel particularly pleased about it. But as soon as I develop a toothache I begin to feel the resulting pain immediately and forcefully. Such examples could be multiplied. As Schopenhauer puts it 'we are conscious not of the healthiness of our whole body but only of the little place where the shoe pinches' (41).

From the preponderance of the feeling of pain over pleasure Schopenhauer derives the principle of the 'negativity of well-being and happiness' and the 'positivity of pain' in life. He writes: 'Hence it arises that we are not properly conscious of the of the blessings and advantages we actually possess, nor do we prize them, but think of them merely as a matter of course, for they gratify us only negatively by restraining suffering. Only when we have lost them do we become sensible of their value; for the want, the privation, the sorrow, is the positive, communicating itself directly to us.' (Schopenhauer 1977, V.I, 412). This intuition is confirmed by the fact that conventional wisdom is forever reminding us not to forget the good things that we have and to be grateful for our many blessings. Indeed popular Victorian homilies such as 'I complained that I had no shoes until I met a man who had no legs!' express this motto, rather crudely in this case, but quite well.

Schopenhauer argues that our so-called pleasures are often no more than the relief of some want or deficiency which is a form of pain. Thus eating and drinking may be pleasurable but basically they satisfy hunger or thirst. Thus it is more a mitigation of pain rather than a gain of pleasure as such. Furthermore, pleasure is usually short-lived. Gratification ends the pleasure and we are then beset by other wants and desires. For 'the will, of which human life, like every phenomenon, is the objectification, is a striving without aim or end' (414). Hence it is impossible to attain lasting satisfaction. Moreover the gratification of our wish or desire rarely matches our expectation and thus often brings disappointment in its train (411-3).

However one exception to this is aesthetic pleasure. It is the only kind of pleasure that Schopenhauer finds unrelated to willing and striving. It is in the pure contemplation of nature, e.g. a sunset, or in the presence of artistic creations such as painting and music that we are temporarily released from our slavery to the will, and the beautiful becomes truly pleasurable. However this pleasure is available to only a few. As Schopenhauer explains: 'For that which we might otherwise call the most beautiful part of life, its purest joy, if it were only because it lifts us out of real existence and transforms us into disinterested spectators of it – that is pure knowledge, which is foreign to all willing, the pleasure of the beautiful, the true delight in art – this is granted only to a very few, because it demands rare talents, and to these few only as a passing dream' (405). For these moments do not last long and soon willing and striving resumes its hold on us. In any case the vast majority of people do not have the capacity to enjoy intellectual pleasures. Schopenhauer does not consider popular forms of entertainment and pastimes as a substitute for aesthetic pleasures. Here he shows himself to be an elitist unwilling to grant the masses reprieve from willing and absorption into the spectacle before them, e.g. at a sporting event, the circus or music-hall, in a manner paralleling the appreciation of arts (405-6). Indeed he believes that sports, card playing and similar pastimes are simply a means to stave off boredom.

Boredom: If willing and striving form one pole of our existence the other is boredom or ennui. This is something that affects the higher animals and of course humans in particular. 'Ennui', says Schopenhauer, 'is by no means an evil to be lightly esteemed'. For 'as soon as want and suffering permit rest to a man, ennui is at once so near that he necessarily requires diversion. The striving after existence is what occupies all living things and maintains them in motion. But when existence is assured, then they know not what to do with it; thus the second thing that sets them in motion is the effort to get free from the burden of existence, to make it cease to be felt, "to kill time", i.e. escape from ennui' (404). Indeed for Schopenhauer, this is "a consequence of the fact that life has no *genuine intrinsic worth* , but is kept in *motion* merely by want and illusion. But as soon as this comes to a standstill, the utter barrenness and emptiness of existence becomes apparent." (quoted in Foster 1999, 216). This may be putting it too strongly but undoubtedly the consciousness of our existence wears heavily on us if we are not engaged in some physical or mental activity. Hence, finds Schopenhauer, public authorities are everywhere conscious of this evil and make every effort to provide diversions and entertainments to occupy the multitude (Schopenhauer 1977, v. I, 404). Clearly the experience of boredom is a price we have to pay for the fact of our self-consciousness including the awareness of time and our suspension in it.

Humans vs. Animals: Humans suffer more than other animals for a number of reasons. Animals have few needs and when these are met they are contented. Moreover they live in the present and have no sense of time - no sense of the past or the future and above all no anticipation of death. Not so with man. First, our desires and wants are far greater and therefore our disappointments are keener. Whilst we are capable of enjoying many more pleasures than the animals – ranging from simple conversation and laughter to refined aesthetic pleasures - we are also far more sensitive to pain. We not only suffer life's evils but unlike animals are conscious of them as such and suffer doubly on that account. Most importantly perhaps it is our consciousness of temporality that makes us suffer the anxieties and fears of accidents, illnesses and the knowledge of our eventual decay and death. The idea of our disappearance from the world as unique individuals is a matter of great anguish and makes us look for all kinds of means of 'ensuring' our immortality. In the main it is religious beliefs that cater to this need. As a professed atheist Schopenhauer finds these and many other aspects of religion as mere fables and fairy tales , a means of escaping the truth about existence including our utter annihilation as individuals by death.

The terrors of existence haunt humans alone, not plants and animals. Moreover death brings us face to face with the vanity of existence. '*Time* and that *perishability* of all things existing in time that time itself brings about Is simply the form under which the will to live ... reveals to itself the vanity of its striving'. (Schopenhauer 1970, 51). Indeed that 'the most perfect manifestation of the will to live represented by the human organism, with its incomparably ingenious and complicated machinery, must crumble to dust and its whole essence and all its striving be palpably given over at last to annihilation – this is nature's unambiguous declaration that all the striving of this will is essentially vain.' (54).

Death and the transitoriness of all things lead humans to question the very nature of their existence. Thus 'To our amazement we suddenly exist, after having for millennia not existed; in a short while we will again not exist, also for countless millennia'(51). This does not make sense for it makes our birth as well as our death, in short life itself, an entirely contingent affair. It therefore raises the question what is it all about. With all the sufferings that human beings have to undergo, with all the effort that they have to expend in the struggle for survival, the cruelty and injustices that they sees all around them and with death as the inevitable end the pointlessness of existence to which they are called, and programmed to continue via reproduction, seems nothing short of a monstrosity. Of course 'the futility and fruitlessness of the struggle of the whole phenomenon (of existence) are more readily grasped in the simple and easily observable life of animals' (Schopenhauer 1969, v.II, 354). The effort and ingenuity they expend in survival and reproduction 'contrast clearly with the absence of any lasting final aim' (354). And the same is true of humans. Despite the elaborate superstructure of civilization that they have built around life the basis of their existence remains the same as that of other animals. It consists of maintaining one's existence and reproducing the species. We are as much nature's dupes as are other living creatures with however one difference. We have the possibility of denying the will-to-live which keeps us in bondage to nature and subjects us to the futility of existence and its continuation through reproduction.

The Ascendancy of Evil over Good: We should note that Schopenhauer does not deny the presence of good beside evil in human existence. However, for him no amount of good can wipe off the presence of evil in the world. "It is quite superfluous to dispute', says Schopenhauer, "whether there is more good or evil in the world, for the mere existence of evil decides the matter, since evil can never be wiped off ... by the good that exists along with it or after it" (quoted in Janaway 1999a, 332). Clearly this involves a moral judgment. Instead of a utilitarian calculus of good and evil, Schopenhauer here takes an absolutist ethical standpoint. However he is not consistent for he also argues that if we could draw up a list of various kinds of sufferings that a person could be subjected to – as we know from history and of everyday happenings around us - and compare it with all the possible pleasure and happiness that he might receive from life, again realistically, there can be little doubt about how the balance would tilt. In some ways Schopenhauer's argument here is similar to that of the preponderance of pain over pleasure as we saw above. We appreciate the good that exists in the world, e.g. genuine compassion for others, acts of kindness, care and concern for the weak and vulnerable, altruism as opposed to egoism. However the intensity and the trauma involved in murder, rape, torture, wanton acts of cruelty cannot be matched by acts of goodness which scarcely register with the same force.

In his judgment of the world as a whole Schopenhauer finds the overwhelming preponderance of evil, which is fundamental to existence, only slightly mitigated by the good which is a marginal and subsidiary element of human life. Hence concludes Schopenhauer, 'Life is a business, whose returns are far from covering the cost' (Schopenhauer 1969 v.II, 353). Schopenhauer tends to repeat statements, such as 'the game is not worth the candle' which betokens a crude utilitarianism, a cost-benefit approach to life. In fact the overall impression conveyed by his philosophy is that of a moral indictment and a metaphysical rejection of the world, based on the nature of the blind will underlying existence. And it is not surprising that in spite of his uncompromising atheism there is an evident streak of a religious perspective on life. This is evident from his use of the vocabulary of 'sin', 'redemption', 'salvation' and the like. Schopenhauer finds three of the world religions, viz. Brahmanism, Buddhism, and New Testament Christianity, sympathetic to his view of life. What is common to these and what they share with his world-view is 'pessimism', i.e. a negative view of worldly existence, and the search for salvation from it. By contrast he finds Judaism and Islam to be 'optimistic' religions (605, 623). They exalt earthly existence, affirm the will-to-live and thus perpetuate the pain, suffering and misery of existence. Indeed he finds all forms of optimistic doctrines pernicious and 'wicked' in that they turn a blind eye to the sufferings of all living beings and perpetuate the illusion of life as 'good'(Schopenhauer 1969 V. I, 325-6)

Schopenhauer believes that religions communicate what they hold to be the truth about existence in the form of myths and fables which the masses can understand. Philosophy, on the other hand, presents its view of existence in an abstract and conceptual manner which is only comprehensible to a small minority of educated people. Nonetheless they are but two forms in which the metaphysics of existence is expressed. Schopenhauer offers an ingenious interpretation of Christianity in line with his own philosophical viewpoint. For him 'The doctrine of original sin (affirmation of the will) and salvation (denial of the will) is really the great truth which constitutes the kernel of Christianity' (405). In the Fall of Adam Christianity symbolizes man's affirmation of the will-to-live. His sin bequeathed to us manifests itself in time through the bond of generation, causing us all to partake of suffering and eternal death. On the other hand Christ, God become man, symbolizes salvation through the denial of the will.

Sex, Reproduction and Guilt: The stress on celibacy in the philosophies of salvation, whether religious or secular, points to the importance of sexuality in the reproduction of the species and the bondage of man to nature. Not surprisingly, the sexual impulse and the intense pleasure in the act of coitus, though momentary, represent the most direct and powerful manifestation of the will-to-life. For what is involved here is no less than the continuation of the species for which nature has programmed us. And it is the blind urge to exist and to propagate that stupefies us into accepting the illusion that to be a human individual is worthwhile (Janaway 1999, 1). Thus something that ought not to be continues its existence driven by the will-to-life. Put differently man's life takes the form of a 'compulsory service that he is in duty bound to carry out. 'But who has contracted this debt? His begetter, in the enjoyment of sexual pleasure'. Because 'the one has enjoyed this pleasure, the other must live, suffer and die' (Schopenhauer 1969, V.II, 568).

In spite of the importance, indeed centrality, of sex in our lives Schopenhauer finds a conspiracy of silence around it. It is ever-present in our lives yet never mentioned. And philosophers too with rare exceptions have ignored the phenomenon altogether. Indeed sexuality in general, and the genitals and sexual intercourse in particular, are associated with shame and guilt. Thus we find that the act through which the will affirms itself and humans come into existence is one of which people are ashamed of and 'which therefore they carefully conceal; in fact if they are caught in the act, they are as alarmed as if they had been detected in a crime' (569). Upon cool reflection we often think of coitus 'with repugnance, and in an exalted mood with disgust' (569). Why, wonders Schopenhauer, this guilt and shame at this strongest expression of the will-to-life? Clearly if our existence was a 'gift of goodness', a praiseworthy and commendable state, the act that perpetrates it would have had a different complexion. On the other hand if it was a wrongdoing, an error, a false step as it were to which we were compelled by blind will then we should feel exactly as we do about this act, i.e. with guilt and shame. 'No wonder not only coitus but the body parts that serve procreation are treated with shame' (570). And it is significant of nature's symbolism that the individual makes his entry in this world 'through the portal of the sex organs' (571).

Romantic love, which creates the illusion of being something sublime and exalted, is at bottom but a ruse of nature to bring men and women together for the purpose of reproduction and the continuation of the species. For the 'consummation' of love requires the physical union of the two lovers and in the child born they see their love sealed physically as it were. But there is more to it than that. We find that the glances of these lovers 'meet longingly' yet ever so 'secretly, fearfully and stealthily'. Why? Because in their heart of hearts they realize that they 'are the traitors who seek to perpetuate the whole want and drudgery, which would otherwise speedily reach an end; this they wish to frustrate, as others like them have frustrated it before' (Schopenhauer 1977, V.III, 375).

Denial of the Will and Liberation: Not surprisingly Schopenhauer comes to the conclusion that 'We have not to be pleased but rather sorry about the existence of the world, that its non-existence would be preferable to its existence: that it is something which at bottom ought not to be' (Janaway 1999a, 332). Clearly the pain and suffering and other negative states of being that humans and other living beings are subjected to, including that which they inflict on each other, and the ultimate futility of existence are what leads Schopenhauer to the above conclusion. The problem then is how it might be brought to an end, i.e. how to escape from the bondage of the will and all that it entails. Schopenhauer is not altogether pessimistic about our prospect of salvation. Not unlike Brahmanism and Buddhism he too believes that knowledge is the key to our liberation from existence. What is required is the denial of the will-to-life.

However, nature has implanted a strong will-to-life in us which stubbornly resists denial. And the less developed their intellect and consciousness the more are people beholden to the will. Thus the 'lower a man stands in an intellectual regard' the less is existence itself a problem for him; 'everything, how it is and that it is, appears to him rather a matter of course' (Schopenhauer 1977, V. 1., 360). His intellect remains 'perfectly true to its original destiny' which is to serve the will. It thus remains 'closely bound up with world and nature, as an integral part of them'. (360). However even for those few who have been able to see through the veil of Maya, so to speak, and have grasped the true nature of existence denial of the will-to-life presents a formidable challenge. Consciousness and intellect have to wage a constant battle against the natural pull of the will (505-6). Nonetheless liberation can only come through the denial of the will.

But how can this be achieved? It is asceticism and the renunciation of all desires and striving that can lead to the renunciation of the will. The highest expression of such willlessness is to be found among saints and other noble souls who attain to a state of holiness. History as well as art provides us with many examples of such individuals who may be Christians, Hindus or Buddhists, i.e. belong to religions of salvation, or have no religious affiliation whatsoever. What they have in common is that state of mind in which renunciation of the will becomes possible. The route to this is through knowledge which may be abstract and philosophical or intuitive and spiritual. In either case it involves an understanding which sees through the surface reality or the phenomenal view of things and grasps the essence of existence as a blind force of will, a force that dominates the life of all creatures. The process of enlightenment might begin with the loosening of the sense of egoism and individuation which form an integral part of the will. The result is a realization that the same will-to-life is present in all creatures. Echoing Lord Krishna in the Gita (see Ch. 1 above) Schopenhauer writes that a man 'who recognizes in all beings his own inmost and true self, must also regard the infinite sufferings of all beings as his own and takes on himself the pain of the whole world' (489). He is no longer concerned with the changing joys and sorrows of his own person. Rather having seen 'through the principle of individuation, all lies equally near him. He knows the whole, comprehends its nature, and finds that it consists in a constant passing away, vain striving, inward conflict, and continual suffering (489)'. After such knowledge why should he assert life any more through acts of will? Unlike those still in thrall of egoism and individuation, which provide them with *motives* for volition he has none. His knowledge of the whole, the nature of the thing-in-itself, 'becomes a *quieter* of all and every volition' (489). The will turns away from life and shudders at the pleasures it recognizes as the assertion of life. He 'now attains to the state of voluntary renunciation, resignation, true indifference, and perfect willlessness' (490). He desires no sensual gratification and denies the sexual impulse totally. He disowns this nature which appears in him already expressed through his body.

Thus 'Voluntary and complete chastity is the first step in asceticism or the denial of the will to live.' Quite apart from renouncing sensual pleasure this act is important in that it denies 'the assertion of the will which extends beyond the individual life' through procreation and ensures that with the life of the body the will also ceases (491). Asceticism also involves voluntary and intentional poverty, the giving away of possessions and resources, as a means of mortifying the will 'so that the satisfaction of the wishes, the sweet of life, shall not again arouse the will, against which self-knowledge has conceived such a horror'(493). Although the ascetic's bodily existence, as the manifestation of the will, shall continue the individual will nourish the body sparingly lest its vigor and well-being ignites the will. He will accept all insults and wrongs returning good for evil. He will break down the will through constant privation and suffering so that when death comes it will merely put an end to a weak residue of the manifestation of the will which has long since perished through free-denial of itself (493). Such is Schopenhauer's view of the denial of the will to live, the thorny path to achieve liberation.

Schopenhauer compares the state of willlessness of such a person with the feeling when aesthetic pleasure, the enjoyment of the beautiful, silences our will temporarily and lifts us above all wishes and cares of the world. As he puts it 'we become, as it were, freed from ourselves' (504). This temporary state of liberation and tranquility is enjoyed continually by the individual who 'after many bitter struggles with his own nature' has at last overcome the will-to-life. In Schopenhauer's lyrical prose, he 'continues to exist only as a pure, knowing being, the undimmed mirror of the world. Nothing can trouble him more, nothing can move him, for he has cut all the thousand cords of will which hold us bound to the world' (504-5). 'Life and its forms now pass before him as a fleeting illusion, as a light morning dream before half-waking eyes, the real world already shining through it so that it can no longer deceive; and like this morning dream, they finally vanish altogether without any violent transition' (505). In many ways this is reminiscent of Buddhist nirvana or Hindu moksha experienced in this life.

This path to liberation is arrived at through knowledge - intuitive or otherwise - about the world as suffering. There is however another way in which the will-to-life is negated and that is through suffering inflicted by fate. It is through extreme suffering experienced personally rather than through knowledge that results in resignation and the virtual extinction of the will. This, Schopenhauer believes, is the more common way to willlessness than the path of knowledge. For the latter not only involves identification with the world's sufferings but going beyond it. And only in a few cases is this knowledge sufficient to bring about the denial of the will. Why? Because even with an individual 'who approaches this point' the blandishments of the will are ever present and act as a 'constant hindrance to the denial of the will, and a constant temptation to the renewed assertion of it'. Hence in most cases the will must be broken by great personal suffering 'before its self-conquest appears' (507).

It is only 'when suffering assumes the form of pure knowledge, and then this knowledge, as a *quieter of the will* , produces true resignation' that it can be the path to salvation' (397). Indeed true salvation, 'deliverance from life and suffering, cannot even be imagined without complete denial of the will. Till then, everyone is nothing but this will itself' (397). The difference between the two routes to salvation lies not in the end state, which is identical, but how it is arrived at, i.e. through knowledge of the sufferings of the world or through personal experience of suffering.

Rejection of Suicide: Significantly enough, Schopenhauer rejects suicide as a path to the denial of the will. Talking of suicide in general he claims that far from denial it is in fact an affirmation of life. For what suicide involves is not turning against the will to live but rather expressing some great disappointment or dissatisfaction with life. The people who commit suicide will life. They desire pleasure and happiness, but find that their own life has not given them what they wanted and therefore choose to end it. This is indirectly an affirmation of the will. The same could be said about people who seek to escape pain and suffering by ending their life.

The essential difference between denial of the will and its affirmation through suicide is that whereas the former involves shunning the *pleasures* of life the latter involves shunning its *sorrows*. Personal escape from pain and suffering, even in the extreme form of self-destruction, does not amount to denial of the will. Salvation cannot be achieved without having the knowledge or understanding about the nature of the will to live and a conscious renunciation of this will with all that it entails. Suicide does not meet this condition for the suicide has not conquered 'his own nature'. And further the suicide merely denies 'the individual, not the species....it is a quite futile and foolish act, for the thing-in-itself remains unaffected by it' (Schopenhauer 1969, V.1, 399). But what, we may ask, of metaphysical suicide? For arguably if I come to understand the true nature of phenomenal existence, i.e. as the outward manifestation of a blind will to live which involves endless striving and suffering without any ultimate aim or purpose and of which I am a part, it is almost a moral obligation on my part to dissociate myself from it. And what better way to do this than to commit suicide and end this individual manifestation of the will? But this is not a line of argument that Schopenhauer would accept. Although he does not discuss metaphysical suicide per se it would seem that the same argument applies here as to suicide in general. Thus it could be construed as an act of egoism which seeks to end the phenomenal manifestation of the will without really expelling from within the will to live and attaining true willlessness.

It seems that Schopenhauer's notion of the denial of the will demands that we continue our bodily existence while at the same time denying the will. This requires complete freedom from egoism, the practice of asceticism, self-mortification, celibacy and the like. The goal seems to be to attain a state of knowledge and consciousness while denying all willing. What remains is only a weak residue of life which disappears with death. Schopenhauer emphasizes that this is not like other deaths because here 'the inner nature itself is abolished' (Schopenhauer 1977, V.1, 494). Schopenhauer scholars such as Dale

Jacquette have noted his 'enigmatic remark' in this connection, viz. "what everyone *wills* in his innermost being, that must he *be;* and what everyone *is*, is just what he *wills*."(quoted in Janaway 1999a, 308). This is reminiscent of Buddhism and Brahmanism where there is an insistence on being free of all desires and attaining an inner detachment from the world. Failing this one cannot achieve liberation and will be reborn. Presumably for the same reason suicide is not a permanent way out of existence for one's inner soul has not been free of attachment to the world and therefore rebirth will follow.

However the main difference between the Schopenhauerian notion of salvation, i.e. the denial of the will to live, and these religious conceptions of moksha and nirvana is that the two latter are based on belief in the cycle of rebirths from which liberation is sought. This requires a total freedom and inner liberation from the world which comes through a form of gnosis arrived at through a long drawn process of asceticism and associated practices. Schopenhauer's secular philosophy does not involve belief in rebirth or the existence of a 'soul' distinct from the body. Yet his notion of salvation seems to suggest a quasi-religious inner freedom from willing arrived at through a somewhat similar process. What also remains unclear is why in a secular context 'abstract' knowledge about the nature of the will and the decision to end one's phenomenal existence should not be a legitimate way of denying the will to live. Schopenhauer scholars have of course noted the rather over-subtle and seemingly contradictory aspects of his conception of the self as well as the will, both of which are involved in the denial of the will to live (see e.g. Jacquette 1999, 306-10; Janaway 1999a, 335-40).

<u>Summary and Comments</u>: Schopenhauer's view of existence may be said to be broadly similar to that of Brahmanism and Buddhism. In common with these religions Schopenhauer finds human existence to be an 'evil' – full of pain and suffering, ending in death and with no aim or purpose other than its own perpetuation. He relates life's suffering to his metaphysics of will – a blind urge or will to live which drives all living beings including humans to survive and reproduce themselves. The ensuing struggle for existence pits individuals of the same species against one another as well as one group or species against another resulting in conflict, violence, pain and suffering. In common with Brahmanism and Buddhism, Schopenhauer too believes that human willing and striving, in short desires, is endless and can never be satisfied. Thus through these innate drives a pointless existence, with its pain and suffering and insatiable striving, is perpetuated. An important insight of Schopenhauer is that life thrives on devouring other lives. In the animal world, whether on land or in the sea killing and eating other living beings goes on quite openly. In the human world the killing is systematic and relentless but hidden from view. Nonetheless humans systematically kill and eat animals, fish, plants and other forms of living things in order to survive. In this sense killing and inflicting suffering is intrinsic to life. The cruelty involved in existence is most clearly evident in the state of nature but in the human world too history and everyday living provides ample evidence of the same phenomenon. Although good also exists it is marginal and weak compared with the ferocious and dominating presence of evil in its multifarious forms. Moreover for Schopenhauer the very presence of evil condemns existence and no

evil act can be undone by any previous or subsequent good action.

Comparing humans with animals Schopenhauer finds that while the former are capable of enjoying many refined pleasures they are also subject to more pain and suffering. For example humans are aware of the finitude of life, of inevitable aging and decay and the painful and protracted wait for death which comes as the end. It is above all one's eventual death and disappearance that underlies the vanity of existence. However the will to live implanted within us creates an innate bias in favor of survival and reproduction. But moral sensitivity and an impartial view of what we are and what our lives entail should make it quite obvious that life in general and human life in particular is a kind of aberration, an error which ought not to be. In common with Brahmanism and Buddhism Schopenhauer believes that real knowledge about the nature of the world, a world driven by the blind and insatiable will to live, is the key to emancipation from existence.

The brutes, animals without consciousness, cannot throw off the shackles of nature. But man through his intellect and self-awareness has the capacity to look at his own existence as it were from outside and free himself from bondage to nature. To achieve emancipation, i.e. ultimate freedom from the grip of the will to live, it is necessary to practice asceticism including celibacy which leads to freedom from all worldly desires and the renunciation of the world. Thus purged of the will to live one can meet death with perfect calm, resignation and indifference secure in the knowledge that one has attained a state of willlessness, and death only ends the physical residue of the manifestation of the will. However Schopenhauer realizes that the struggle to achieve willlessness is a difficult one as the presence of the will within us always tempts us to seek pleasure and to avoid pain. Yet to be an ascetic and a world-renouncer one must avoid pleasure and embrace pain and suffering. Only a small minority of individuals seem capable of attaining the knowledge that liberates and the ascetic practices that lead to the conquest of the will. But the lives of saints and ascetics of many lands and of many beliefs and faiths show what human beings can achieve. Their holiness, unworldliness and state of blissfulness stand as shining examples of liberation from the quagmire of existence. Here Schopenhauer's conception of the transcendence of the will comes very close to the Hindu concept of Moksha and the Buddhist notion of Nirvana. It is a state of beatitude that is beyond comprehension and beyond description in terms of subject-object distinction, for it lies beyond these dualities. From our worldly point of view it appears as a void or nothing but from the viewpoint of those who have freed themselves from the will, this our world appears as nothing. Such is Schopenhauer's world-view in a nutshell. What we offer below are a few critical comments and observations on his vision of existence and the road to liberation, the focus of our study.

On the evils of existence: Schopenhauer emphasizes pain and suffering as the defining characteristics of life. And although the pointlessness and futility of existence is a part of his conception of this 'evil' it receives far less attention and emphasis. It is the moral rejection of the world that he emphasizes rather than its metaphysical worthlessness as a contingent phenomenon. Schopenhauer analyses and discusses human sexuality at great length and with impressive insight. An important omission, however, is the suffering of young humans through a long period of sexual deprivation and frustration, a form of suffering, which animals do not have to undergo. Although he lambasts 'optimism' for its disregard of the glaring evidence of gratuitous suffering all around us he fails to emphasize that all procreators and existents are directly or indirectly responsible for the world's evils and its perpetuation. And although Schopenhauer expresses strong anti-natalist views he stops short of considering, not to say advocating, anti-natalism as a means of liberating those still unborn from existence. In other words the idea of *prevention* is almost entirely missing from his notion of liberation which is focused exclusively on individual salvation for those already here. While he notes in passing that non-procreation by all would mean the disappearance of the human race, a prospect that he should clearly welcome, he does not consider it as an act that would at least *in part* deny the will to live by way of abstaining from procreation. Against this it might be argued that Schopenhauer's philosophy is essentially about explaining the world rather than changing it. In short it is largely descriptive rather than prescriptive although his idea of salvation, i.e. the denial of the will and its methodology, takes him into the realm of practice. According to his metaphysics of the will, salvation can only come from the *total* denial of the will which requires extremes of asceticism and self-mortification in order to break down the will to live. That is why he also rules out metaphysical suicide as a form of denial of existence and the triumph of the intellect over the will. For Schopenhauer this act of self-destruction does not extirpate the will to live but simply its physical embodiment. In fact it would count as an act of the affirmation of the will. As we shall see below this somewhat narrow and dogmatic view of salvation, which undoubtedly stems from his metaphysics of the will, is also contradictory.

Schopenhauer equates the will, the all-pervasive thing-in-itself, with the will to live. In short, all willing is affirmation of life. Yet the ascetic or the world-renouncer implicitly *wills* the denial of the will to live which is obviously incoherent. Schopenhauer tries to resolve this problem by suggesting that the ascetic cannot deny the will through volition. Rather with real knowledge of the nature of the will and with ascetic practices the individual reaches a stage when the will simply turns against itself, it 'denies itself'. As Janaway points out, if all willing is will to live then the ascetic cannot *will* the denial of the will to live. And this is the main position of Schopenhauer (Janaway 1994, 95). Indeed Schopenhauer compares his idea of salvation with the Christian notion of grace, i.e. something that comes to the individual from outside. The denial of the will, he writes, 'comes suddenly, as if flying in from without' which the Church calls 'the effect of grace' (Schopenhauer 1969, V. I., 408). This is a strange quasi-religious notion of salvation coming from an atheistic philosopher! However elsewhere he writes that the ascetic has to wage a constant struggle against the will to live which is always seeking to affirm itself within him (Janaway 1994, 95).

Thus Schopenhauer's portrayal of the beatific state of willlessness comes up against his notion of the constant struggle that the ascetic has to wage in order to conquer the will and maintain the state of willlessness as long as he lives. In this regard his notion of salvation cannot compare with the Buddhist nirvana which is akin to a mystical state of transcendence and tranquility. Many Schopenhauer scholars have commented on his fuzzy and incoherent notion of the state of willlessness, both in respect of its nature and how it is arrived at (Navia 1980, 178-9; Janaway 1994, 94-5).

Overall Schopenhauer's view of existence – the fundamental pointlessness of life with its endless willing and striving and which involves much pain and suffering – has a great deal of coherence and validity. In the last analysis, it amounts to a moral and metaphysical rejection of life. On the other hand, as we argue below - and this point has been often noted - his notion of liberation from the shackles of the will to live remains unconvincing and unsatisfactory in the extreme. The main reasons are as follows.

Paths to Liberation: According to Schopenhauer there are two paths to liberation. The first, arrived at through a real understanding of the nature of the will to live, can lead to its complete denial. Such knowledge has to be followed by extreme ascetic practices in order to break down and overcome the will. However, only a few people are capable of attaining this insight or knowledge and even fewer that of denying the will. Saints, holy men and mendicants are prime examples of those who achieve liberation in this way.

The second and, according to him more common, route to salvation is through the breakdown of the will as a result of an inordinate amount of suffering. Many individuals, having suffered a lot and lost out in life, suddenly come to realize the vanity of existence and of all striving and willing and deny the will to live. However Schopenhauer is hard put to find credible examples of such people. One of these is Gretchen, a fictional character, in Goethe's Faust.

Others he mentions are condemned criminals waiting execution, noblemen, adventurers and kings. As Nietzsche and others have observed, the idea of salvation resulting from extremes of suffering and despair is nothing short of grotesque (Janaway 1999a, 341). In any case Schopenhauer's examples are somewhat exceptional and generally speaking extremes of suffering do not necessarily lead to resignation and loss of the will to live. Suicide or a life of neglect and apathy could well be the more likely outcome. But the latter does not connote Schopenhauerian willlessness. In any case for the vast majority of humankind Schopenhauer's philosophy seems to offer no hope of liberation. They must continue to be driven by the will to live and remain in bondage to nature.

A further difficulty with Schopenhauer's idea of liberation, stemming from his deterministic philosophy of the will, is the question of choice. As we saw above the will to live cannot be denied intentionally. Given the right conditions it has to 'happen' as it were. As commentators on Schopenhauer have noted the entire subject of how one arrives at the right knowledge about the will and translates this knowledge into the denial of the will is shot through with ambiguities and complexities. But one thing is clear. The individual is not free to *choose* to deny the will. All of this is in contrast with the logical approach of Buddhism where the road to salvation and ultimate nirvana is laid out systematically as the eight-fold path consisting of right knowledge and right action. And it is open in principle to all individuals who decide to embark on this difficult road to emancipation. Here the responsibility lies with the individual and emancipation is something earned through merit rather than received as 'grace'. True, in both cases all willing and striving ceases, the will to live is extinguished and the liberated individual attains something like a mystical state of being. In both cases the end state might be the same but who gets there and how seems rather convoluted, and also contradictory in the case of Schopenhauer.

As we pointed out earlier, his conception of the self and the will is overly subtle and self-contradictory and this is a part of the problem (Janaway 1994, 94-5, Chs. 3 and 4 passim.). He has to resort to quite convoluted arguments to reconcile these contradictions and inconsistencies. The mystical sense of 'oneness' with all existence and living beings felt by the will-less is reminiscent of the Hindu conception of moksha where the atman or the individual self recognizes itself as identical with Brahman or the cosmic order. Indeed Schopenhauer uses the term 'god' to indicate the state of being after denial of the will which cannot be expressed in words because we have no conception of that state of being. Thus Schopenhauer's conception of the state of salvation, in common with Brahmanism and Buddhism, turns out to be quasi-religious, It also invokes a mystical state rather than a void or nothingness, pure and simple.

It is interesting to note another element of similarity among all three. For both Hinduism and Brahmanism, which believe in the transmigration of the soul or the inner self, birth as a human being is a great opportunity to liberate the self from entrapment in the eternal cycle of birth and rebirth. Similarly according to Schopenhauer human beings possess the advantage over other forms of life of a more developed intellect and consciousness which enables them to see through the ruse of nature and   free themselves from bondage to nature. He writes, "nothing else can be stated as the aim of our existence except the knowledge that it would be better for us not to exist" (quoted in Janaway 1999, 13). Elsewhere he remarks, 'the value of life lies precisely in this, that it teaches him (man) not to want it' (Schopenhauer 1970, 65). Yet as he admits only a few are capable of attaining to the necessary knowledge and fewer still to succeed in freeing themselves from the will to live. In fact Schopenhauer's path to liberation is so thorny, so lofty, so extreme and also so fortuitous (the will cannot be denied by intention but has to deny itself through 'grace') that very few are willing or capable of following such a path.

There are two issues here. First, Schopenhauer believes in the *total* denial of the will which leaves no scope for a *partial* denial of the will. As we have noted above, future generations could be spared the sufferings of existence through *prevention*, i.e. non-procreation. This does not require those extremes of asceticism and self-mortification in order to break down the will. As Schopenhauer himself remarks, 'If the act of procreation were neither the outcome of a desire nor accompanied by feelings of pleasure, but a matter to be decided on the basis of purely rational considerations, is it likely that the human race would still exist? Would each of us not rather have felt so much pity for the coming generation as to prefer to spare it the burden of existence, or at least not wish to take it upon himself to impose that burden upon it in cold blood?' (Schopenhauer 1970, 47-8). However we have to acknowledge that in Schopenhauer's time contraception was scarcely developed and it was almost impossible to break the bond between coitus and procreation. It is also true that Schopenhauer puts a great deal of emphasis on total sexual abstinence as a part of asceticism, which leads to the denial of the will. But refraining from procreation as such is not a course of action suggested by him. Yet it is an important part of the denial of the will, being one of the strongest expressions of the will to live. As we shall see later, in the present day context Benatar argues the case for abstaining from procreation as an eminently desirable and feasible course of action available to all which would spare future generations from being brought into existence. It should also be noted that in common with Brahmanism and Buddhism, Schopenhauer's philosophy is also concerned with the salvation of *existing individuals* through attaining willlessness rather than saving future generations from the pain and suffering of existence. It is a reasonable assumption that the two routes to salvation proposed by Schopenhauer are likely to involve mature individuals who most likely have already created progeny. Thus their own salvation and denial of the will may be seen as compromised - a Pyrrhic victory over the will - since they have already colluded with nature in bringing to life new existents and new victims of suffering. This is a point to which Schopenhauer pays no attention.

There is another issue that deserves notice. Schopenhauer claims that man's intelligence 'is already sufficient for imparting to the will that knowledge in consequence of which the will denies and abolishes itself' (Schopenhauer 1969, V. II., 610). Elsewhere he writes that those with a lower level of intellectual development are less likely to see through the real nature of existence, i.e. their intellect still remains in service of the will (610). The implication surely must be that with the spread of education in future years and with more and more people receiving higher education intelligence will become more autonomous of the will. Thus more and more people will seek to liberate themselves and future generations from bondage to nature. In fact Schopenhauer makes several references to the intellect and will as a source of potential conflict. As Janaway (1999, 5) remarks, the central thought of Schopenhauer's work is that 'knowledge culminates in a kind of abnegation' i.e. self-realization results in self-cancellation. While this is clearly true at the individual level the question is whether it goes further than that.

According to Atwell, Schopenhauer implies that human development is leading towards a single purpose, viz. that of the intellect and consciousness becoming increasingly aware of the nature of the will and denying it. This is the meaning of the statement that "nothing else can be stated as the aim of our existence except the knowledge that it would be better for us not to exist" (13). Elsewhere Schopenhauer writes that with greater knowledge intelligence ceases to be a tool of individual nature, i.e. animal nature, and turns away from existence with its 'mere repetition and tedium through endless time'. And finally, and even more clearly, 'the aim of all intelligence can only be a reaction to a will; but since all willing is error, the last work of intelligence is to abolish willing, whose aims and ends it had hitherto served' (Schopenhauer 1969, V. II, 610).

Clearly Schopenhauer's assertion of reason's triumph over will is incoherent in that the will to live is the primary driver of human existence and conflict between will and reason (intelligence) is not a part of Schopenhauer's philosophy. It lacks plausibility. As Janaway remarks, 'does the world as a whole strive in order to reach its own non-existence?' Rather it seems that 'ultimate reality endlessly strives simply to be' (Janaway 1999, 13). Indeed as Schopenhauer himself points out even the enlightened saints and dedicated ascetics have to wage a constant battle in order to overcome the allurements of will. How then does he expect human existence to come to an end as a result of increasing knowledge and intellectual advancement of mankind? This is an important point which needed more detailed consideration. As we shall see later Hartman, whose philosophy is in many ways indebted to Schopenhauer, puts forward the thesis of the growing ascendancy of the intellect over the will in the course of evolution which will eventually lead to humanity's self-annulment. But the conflict between will and reason is not something Schopenhauer cares to dwell on or develop. Since his philosophy has no place for history or human development this important issue relevant to the future of liberation remains unexplored. We shall take up this point later. For the moment we conclude this section by reiterating the point that Schopenhauer's notion of liberation from existence is neither clear nor consistent. Moreover its practicality seems extremely limited. Does Hartmann have a better answer? We consider Hartmann's philosophy next.

Eduard von Hartmann: Reason over Will

Hartmann's philosophy has many similarities with that of Schopenhauer. For Hartmann, as for Schopenhauer, the pain and suffering involved in life far outweigh its pleasures. And for Hartmann too the evils of existence are fundamental and irremediable. He differs from Schopenhauer in that he looks for a collective rather than individual solution to overcoming existence. Secondly, unlike Schopenhauer he has an historical and evolutionary perspective on the process of world-transcendence (Hartmann 1884, V.III).

Hartmann interprets all living phenomena in terms of an underlying element - the Unconscious. It comprises the will or the non-logical element – the driving force - and reason or idea which is the logical element. With the evolution of the world process the logical element develops and becomes stronger. It is embodied in consciousness, which is most fully developed in man. For Hartmann it is the heightening of consciousness in the course of evolution and the increasing realization of the true nature of existence that is the key to emancipation. However before this stage can be reached humanity, the most conscious part of existence, has to go through a series of 'illusions' which keep it trapped in the cycle of existence. There are three stages or forms of illusions concerning happiness. The first is that well-being and happiness can be achieved in the here and now; the second is that earthly life is inherently evil in nature and can never satisfy but that there is life after death which will bring blissful immortality; the third illusion is that man can achieve mastery over the world and that material progress can lead to a happy and fulfilling existence. In short a better world can be attained in the future (12, 79, 94).

Hartmann finds that humanity has already passed through the two earlier stages (see below) and is now in the third stage. He seeks to demonstrate that all three are illusory and that the quest for happiness and well-being is doomed to frustration and failure. Looking at the history of the Western civilization and treating it as paradigmatic of humanity's evolution, he argues that the idea of happiness in the here and now is dominant at the beginning of civilization and thought. Greece and Rome as well as Judaism represent the illusory quest for fulfillment and happiness in the present. This gave way to the second stage represented by Christianity and the Middle Ages. Here earthly existence is seen as evil - full of pain and suffering and man as 'fallen' and sinful. Earthly life is seen in terms of a preparation for an afterlife with the promise of eternal life and deliverance from evil. This religious belief in a supramundane life of bliss begins to weaken with the Renaissance and Reformation. Skepticism and humanism start making inroads into the notion of heavenly peace and happiness and the focus shifts to this world. The development of science, including that of the social sciences, and technology gives man greater control over the physical and social world. This leads to a belief in the idea of progress and the prospect of a better future for humanity. The result is increasing secularization, the waning of religious beliefs, and a gradual weakening and marginalization of the idea of happiness in a life hereafter. But in the long run this secular belief in a happy life, made possible through material progress, is also doomed to failure. Why? Because progress may diminish but never get rid of the fundamental sufferings of existence, e.g. illness, age and other painful conditions. Second, material progress there may be but not moral progress. Civilization can only change the form but not the substance of the innate egotism and destructive impulses of humankind. Third, the more the 'palpable external ills of human life are removed' the more evident it will be that the source of pain and suffering is existence itself' (114). It is only after humanity has seen through the illusions of progress that it will be ready at last to seek deliverance from the evil of existence (12, 79, 91, 114-5).

What are the main reasons for H's assertion that the ills of existence – the pain and suffering and the sheer immorality of man – far exceed the good and the pleasure that it affords us? The reasons he puts forward are not particularly original or indeed very different from those advanced by Schopenhauer, as we saw earlier. Thus according to Hartmann the so-called good things of life, e.g. health, youth, freedom, material possessions, do not represent a positive value to us. Rather they signify the absence of negativities such as sickness, old age, servitude, and poverty and insecurity. He emphasizes the 'felt' or subjective experience of these and argues, along the lines of Schopenhauer, that whereas we feel these negatives of life keenly we take the positives for granted. For example, having normal eyesight and being able to see does not make us happy. On the other hand we feel the loss of sight and the resulting deprivation far more keenly. In short, the presence of pain, suffering and deprivation are felt far more strongly than their absence. Moreover many of the pleasures of life come at a great cost. For example from the beginning of puberty until the time that a young man can afford to get married he has to endure the agony of intense sexual frustration. He has to seek relief through all kinds of unnatural and abnormal, if not immoral, practices. The pleasures of love, marriage and family also come at a great cost. The pain and suffering involved in pregnancy and childbirth for women, the arduous task of caring for and bringing up children plus all other privations and disappointments involved in family life are scarcely compensated by the pleasures and gratifications that it brings us. Indeed it is instinct and unconscious attachment to life (the element of will) and sexual compulsion that urges people to reproduce. He agrees with Schopenhauer in his view that without the extreme pleasure in the act of procreation (for men?) no one would inflict life upon someone in cold blood. He reviews a long list of human activities and relationships and finds them a greater source of 'toil and trouble' than well-being and happiness (see e.g., 23-5, 47-8). Nonetheless, Hartmann finds that in our assessment of life in general, including the balance of pain and pleasure, we are mostly biased in favor of life. This is primarily due to our instinctive impulse of self-preservation and attachment to life. We call life good not because it is so but because we are instinctively attached to it (8-10) . Hartmann maintains that the more conscious and reflective we are about the nature of our existence the more we can see through life's illusions and appreciate the truth of his viewpoint. It is through education, knowledge and enlightenment of the mind that we become free from the veil of illusion. And thus far this enlightenment has been available to only a small minority of peoples of the world (10).

But how does Hartmann envision the transcendence of existence? The metaphysical principle underlying his philosophy of existence is that once reason or consciousness comes into its own it will see the folly of existence and will seek liberation from its bondage to the will. He believes that the third (the current) stage of human development will gradually raise our consciousness further so that in the long run the greater part of humanity will come to realize the fundamentally flawed nature of existence and its inescapable miseries. Once this perception takes hold, volition will cease and the world process will come to an end. Hartmann sets great store by the growing antagonism between will and reason. The former, he writes, strives after 'absolute satisfaction and felicity' while intelligence emancipates 'itself more and more from the impulse through consciousness' (123,4).

But how precisely would all this come about? Hartmann rejects all individual approaches, e.g. suicide, or the renunciation of the will (Schopenhauer), and looks for a collective solution. Liberation of a few individuals from the yoke of the will means very little since the mass of humanity and the rest of existence will keep going. And Hartmann's objective is nothing short of bringing all of existence to an end. Hence existence must be brought to an end through a collective global resolve. Indeed he envisions not only the end of human existence but of all life forms and their sources to ensure that life and especially human or conscious life may not start all over again. Hartmann puts forward three necessary conditions for the transcendence of existence (135-9). First, human consciousness as a whole should be sufficiently developed to see the folly and futility of existence; second, that the unconscious or the will and spirit (reason) operating in the world should become concentrated in humanity relative to the rest of life [1]; and finally that there should be 'sufficient communication' among the peoples of the world to 'allow for a simultaneous common resolve' (139) in this regard. He finds it difficult to be specific about how exactly the process will unfold. But he believes that in the meantime human development should continue on its course and that it is a moral obligation on everyone's part to contribute to the process of world development until the three necessary conditions are met and existence is brought to an end. For Hartmann the endeavour to bring existence to an end is an *altruistic* act meant to end at the same time the immorality of life as well as its pain and sufferings[2]. Underpinning his belief that world-evolution is moving in that direction is his metaphysical premise that the world began from an original state of rest and non-being and that return to that initial state is immanent in the process of evolution itself. It also stands for the ultimate victory of intelligence and reason over volition and the will-to-live (120-7; Darnoi 1967, 102-3).

Summary and Comments: If Hartmann's view of existence is very similar to Schopenhauer's his perspective on emancipation differs radically from the latter's. And in some respects he makes an original and interesting contribution to the problem. First, unlike Schopenhauer's timeless view of will and existence Hartmann sees the problem of existence in historical and evolutionary terms. This coupled with his idea of a conflict between the will, which seeks to perpetuate existence, and the developing consciousness, which sees through the illusory promise of the will and seeks to put an end to existence, makes the potential for emancipation inherent in the world process itself. Put somewhat differently, this conflict between will and reason develops through the growth of consciousness, most advanced in humankind. And for Hartmann this is what forms the basis for the denial of the will that Schopenhauer speaks of. He is rightly critical of Schopenhauer's notion that the denial of the will occurs through the will turning against itself. This is self-contradictory given that Schopenhauer considers the will as a blind force, with its ceaseless striving and the perpetuation of existence. Certainly Hartmann's idea that the fundamental unconscious, the primordial element behind all existence, consists of both will and idea or reason and that the latter develops in the course of evolution as consciousness offers a plausible basis for a source of growing opposition to the will and its perpetuation of existence. It is worth noting that this is the basis of the belief that 'knowledge' abnegates existence, a belief that is common to all three world-views, viz. Hinduism, Buddhism and that of Schopenhauer, encountered earlier.

The second important idea, linked to the above, concerns the intellectual and cultural development in the course of the world process. This means that as a greater part of humanity develops educationally and intellectually more and more people can become conscious of the fact that pain and suffering are an irremediable part of existence and can see through the veil of illusion which makes us believe in the possibility of progress and happiness. In other words higher intellectual development leads to greater awareness of the evils of existence and the rejection of bondage to nature which perpetuates this evil.

The third point is that although Hartmann's periodization of human history into three distinct phases cannot be considered as having literal validity there is a modicum of truth in this perspective on history. This is especially the case if we look at his third or current phase which he sees as one of increasing secularization, i.e. the waning hold of religion on people's beliefs and world-views. Clearly this provides a fertile ground for the development of a perspective on the world which finds existence morally and metaphysically unacceptable. And although the idea of the greater mass of humanity becoming convinced of the worthlessness of existence seems far-fetched, if not wishful thinking, there is little doubt that a secular world provides a favorable context for the spread of anti-existential ideas and attitudes. At least we can expect this to result in the spread of the philosophy and practice of *anti-natalism* based on the rejection of existence. Indeed as we know, over the last thirty years or so anti-natalism, stemming from a variety of reasons which include, it is safe to assume, anti-existential attitudes as well, has been on the rise. More and more people in the advanced world are choosing not to procreate (see e.g. Basten 2009). Although Hartmann looks towards a collective demise of humanity, indeed of all existence, based on his metaphysical and teleological beliefs we need not take that seriously. The main point is that he is right in drawing attention to the weakening of religious beliefs and the increasing secularization of the world as an important context for the increasing rejection of existence.

Finally, his idea of increasing communication worldwide – a result of economic and technological development – as one of the enabling conditions for greater awareness and action aimed at bringing existence to an end has considerable merit. Leaving aside the fanciful notion of collective suicide by humanity his point regarding increasing communication is an important one, especially taken in conjunction with the two earlier points, viz. greater intellectual and cultural development and secularization. For what we are witnessing today is not only the dissemination of anti-existential ideas through books and journals but also the beginning of a debate on these issues on the internet. To this we can add the formation of associations, e.g. VEHMT (Voluntary extinction of humanity movement), and others which seek to connect people with anti-existential as well as anti-natalist viewpoints[3.] In sum, indirectly more than directly, Hartmann provides us with a set of bold and original ideas on liberation from existence which can be seen as a body of thought which borrows heavily from Schopenhauer but also contributes new ideas relevant to rejectionist theory and practice.

Endnote – Chapter 2.

1. This is a fuzzy notion. However since Hartmann envisions nothing short of the complete cessation of all life he is forced into a position which seeks to equate the end of humans with the end of all life. He seems to conceive of the relative predominance of humans in two ways. First that economic and technological development of the world, including the growth of human population, would advance so far as to marginalize and diminish the rest of nature very considerably, and second that it will so empower humans as to be able to abolish the rest of nature. On closer examination and in light of more than a century of population growth, environmental degradation and the technological advance of humanity, these ideas do not seem quite as fanciful as at first sight. See Hartmann (1884, 135-7).

2. For Hartmann egoism is a major source of immorality and evil in the world. He sees the gradual enlightenment and freedom from the futile quest for pleasure and happiness as the path to overcoming egoism and achieving liberation. For him the Christian notion of immortality is both an expression of, and a major prop to, egoism. He is against all *individual* acts of rejection, e.g. 'willlessness' a la Schopenhauer, suicide, as well as the Buddhist quest for individual nirvana, for they all cater to the self and egoism. His idea of a *collective* resolve on the part of humans to bring existence to an end is based very much on the rejection of egoism in favor of an act of solidarity and transcendence of selfishness. For Hartmann 'morality and religion demand the utter uprooting of selfishness; redemption is not redemption *of* self but redemption *from* self (italics added). The denial of the will is not to be individual in any sense, and nowise partial. It should be universal and final, and should register the extinction of the entire world-process.' (Tsanoff 1931, 329,

326-31).

3. For a list of these organizations see Basten (2009, 15). For VHEMT see <http//:vhemt.org>

Chapter 3

Philosophical Perspectives:
20ᵗʰ century and Beyond

Among the philosophical currents of the 20ᵗʰ century it is existentialism that has the most relevance to the main theme of this book, viz. the implication of man's existence in the world as a conscious being [1] Philosophers and writers such as Heidegger, Sartre, Camus, Unamuno and Cioran are some of the notable figures who may be said to have contributed to this genre of thinking (on the last three writers see Dienstag 2006). However with the exception of Cioran who rejects existence openly and totally, all others, in the end, reconcile themselves to existence on the basis of a variety of rationalizations and viewpoints [2].

Although philosophers such as Heidegger and Sartre analyze man's being-in-the-world in considerable detail they take largely a value-neutral stance on the question of human existence and procreation. Only Camus is explicit in posing the question whether 'life is worth living' and in raising the question of suicide as a response to the 'absurdity', i.e. the meaninglessness, of man's existence. He ends up by arguing in favor of accepting life in spite of its many failings and its fundamental absurdity (Camus 1975). It is worth noting that he is silent on the question of procreation. Heidegger rejects the label of 'existentialist' (in fact only Sartre ever claimed to be one although he too distanced himself from the label later) arguing that his philosophy was concerned with the nature of man's being and not with questions of values, ideologies and other humanistic concerns. Heidegger, Camus and Unamuno were all married and had children. Sartre remained childless but it is not very clear whether it was a rejectionist measure or more a matter of choosing a 'childfree' lifestyle. It appears to have been a bit of both. But apart from his novel Nausea in which the main character pours scorn on the idea of having children, there is little in his writings on the theme of anti-natalism. That leaves Cioran who is said to have claimed that not having children is one of the best things that he has done in his life (Ligotti 2010, 176). As a total rejectionist Cioran's thinking on existence should potentially be of great interest. Unfortunately he is something of a belletrist with an allusive, aphoristic and paradoxical style. His writings take the form of random reflections on a variety of subjects. Philosophical concerns almost always inform his writings yet he remains elusive and unsystematic in the extreme as a writer. It is difficult, if not almost impossible, to grasp his thinking, let alone discern any pattern to it.[3]

Peter Wessel Zapffe: Against Procreation

A little-known Norwegian philosopher and writer, Peter Wessel Zapffe (1899-1990), who may also be considered an existentialist, joins the company of Cioran in being a rejectionist and an anti-natalist. Although Zapffe's magnum opus On the Tragic (1941) is still only available in Norwegian his essay 'The Last Messiah' and other short pieces, including an interview, provide us with an outline of his view of existence. Zapffe is in favor of phasing out human existence through non-procreation or even by a general policy of limiting the number of children begotten below replacement levels (Tangenes 2004). As we shall see below, the South African philosopher David Benatar presents a comprehensive philosophy of anti-natalism as a means of preventing future suffering and the path to liberation from existence. But it is Zapffe who must be credited with being the first rejectionist to come up with the idea of anti-natalism as the way out of existence for humans.[4] It is to Zapffe's main ideas that we turn next.

Zapffe's focus is on the contingent nature of man's existence both at a collective and individual level. Yet the *consciousness* of existence, which makes the human animal unique, creates the need to find an overarching meaning to it all. This is a need that will not go away and yet it cannot be satisfied except through resorting to myths, fables and lies, in short to bad faith. Thus evolution has produced a freak of a species that is conscious of its existence. And that is its undoing, making it fearful of life itself, indeed of its own being. As Zapffe (2004, 2) writes in <u>The Last Messiah,</u> "Despite his new eyes, man was still rooted in matter, his soul spun into it and subordinated to its blind laws. And yet he could see matter as a stranger, compare himself to all phenomena'. Nature having 'performed a miracle with man' later disowned him' (2).

Quite apart from the need for an overarching meaning of life humans have developed values and sentiments absent in all other species. Longing for a just world is one of them, compassion for the suffering of all sentient beings is another. Yet these have no place in the universe. The lack of meaning reveals to man, 'a nightmare of endless repetition, a senseless squander of organic material' while the 'suffering of human billions makes its entrance into him through the gateway of compassion' (2).

Zapffe concludes that human beings are a clear case of a species that 'had been armed too heavily – by spirit made almighty without, but equally a menace to its own well-being'. Here is the 'tragedy of a species becoming unfit for life by over-evolving an ability' (i.e. an over-abundance of consciousness) (2). Besides the knowledge of and fear of death man's creative imagination conjures up new and 'fearful prospects behind the curtain of death' so that even death fails to be the sanctuary that it is meant to be'. Given these inbuilt contradictions and the paradoxical nature of the species Zapffe wonders why it has not gone extinct through the inability to cope with these fundamental conditions of life. His answer is that most people learn to limit the content of their consciousness through a variety of strategies and techniques developed by culture and society. He goes on to elaborate these (3-7).

Essentially they involve the repression of the 'damaging surplus of consciousness'. At least four major types of mechanisms are involved: isolation, anchoring, distraction, and sublimation (4). In practice they may overlap. Isolation is about the 'arbitrary dismissal from consciousness of all disturbing and distressing thought and feeling'. In everyday life this takes the form of a conspiracy or code of mutual silence about the fundamental questions concerning existence.

Anchoring involves a range of institutions, beliefs and social norms which act to protect the self against the consciousness of the void or abyss that surrounds us. Zapffe writes of anchoring 'as a fixation of points within, or construction of walls around the liquid fray of consciousness'(4) . Culture and ideologies provide the main source of anchoring which often act unconsciously but could also be conscious for the individual, e.g. adopting a goal to be pursued, dedicating oneself to a cause. God, the church, the state, the nation are major examples of the resources for anchoring. Distraction is a mode of protection which seeks to deflect attention away 'from the critical bounds' to a variety of impressions. This is most visible and obvious with children who have to be provided with various diversions to stave off the sheer boredom of existence. For grown-ups distraction can take the form of games and entertainments of all sorts. But above all it is the constant pursuit of desires, the continual striving for something or the other that is a major form of distraction. As soon as a goal is reached one moves on to another. Here Zapffe seems to have in mind something like Schopenhauer's idea of constant willing and striving without which we face boredom and the emptiness of existence. But Zapffe argues that this 'striving forward' is equally an 'escape from,' something that is not generally recognized. But escape from what? The answer, expressed in religious terms, is from 'the vale of tears', 'from one's own inendurable condition'. Indeed, writes Zapffe, if 'awareness of this predicament is the deepest stratum of the soul....then it is also understandable why the religious yearning is felt and experienced as fundamental' (7). Sublimation, the fourth remedy, is a matter of transformation rather than repression. Thus stylistic or artistic gifts can transmute existential anxiety and despair into literature or painting which becomes a source of aesthetic appreciation and a form of release from anguish. Of course this particular remedy is available to only a few and therefore may be regarded as the least important of the four methods of coping with the problem of consciousness. Zapffe concludes by claiming that however effective these might have been in the past they cannot save humanity from itself for ever. Writing in the 1930s he notes that communism and psychoanalysis are among the most recent attempts to ensnare the 'critical surplus of cognition'.

He has no doubt that humans will persist in 'dreaming of salvation and affirmation and a new Messiah'. But after many messiahs have come and gone the Last Messiah will come. He will be a man 'who has fathomed life and its cosmic ground, and whose pain is the Earth's collective pain'. He will tell humans the truth about themselves in no uncertain terms. 'The sign of doom is written on your brows'. There is 'only one conquest and one crown, one redemption and one solution'. And that is 'know yourselves – be infertile and let the earth be silent after ye' (9). However Zapffe is under no illusion about how such a message would be received. Led by the guardians of existence, and far more savagely than in the case of the Crucified One, the Last Messiah will be sat upon and torn apart by the multitude,

Zapffe, an assessment: Zapffe has apparently developed these themes at greater length in his book On the Tragic. But the essentials of his viewpoint appear to be as sketched above. Undoubtedly there are problems with Zapffe's argument. The basic premise on which he builds his case for anti-natalism, viz. that human beings cannot bear the burden of consciousness - the over-developed faculty that they are endowed with - can be questioned. (Incidentally his argument is reminiscent of T. S. Eliot's "Human beings cannot bear much reality"). Undoubtedly, man is an anomaly produced by nature and his consciousness of existence gives rise to many metaphysical and existential problems. The need for 'meaning' is one of these and which is at the heart and centre of Zapffe's thought. However as he argues humans have developed a variety of beliefs and institutions, most prominently those centered on religion, to cope with the problem of existence including its legitimation. And if there is a weakening of religious beliefs and increasing secularization a form of secular humanism seems to be taking the place of religion in providing a rationale for existence. The belief in 'progress' driven by science and technology and the prospect of a rising standard of material life for more and more peoples of the earth is a part of this rationale. In any case the protective walls and the many diversions outlined by Zapffe have worked in the past and they could presumably go on working in the future. Moreover one cannot underestimate human capacity to invent new ways of shoring up existence. Thus it is difficult to accept Zapffe's idea that the contradictory or paradoxical state that humans represent will of itself lead to the demise of the species in the manner of the antler with overly large horns - referred to by Zapffe - or other such species that have become extinct. Perhaps Zapffe is speaking here metaphorically. But this is not to suggest that Zapffe's theme of the problem of meaning and its implications lacks validity. Indeed not only the pointlessness of existence and human awareness of the same but also what he calls the 'brotherhood of suffering' of all creatures taken together provide a strong basis for rejecting existence.

But this is more a question of individual enlightenment and choice rather than a self-evident truth which by its very nature calls forth a certain type of action. The 'tragedy' which Zapffe finds in human condition is unlikely to be perceived or felt by more than a minority of individuals, mostly intellectuals. And even among those many are prepared to accept the insoluble paradox of conscious existence, placing 'life' above 'truth'. Not only Nietzsche but also Unamuno and Camus for example come to mind. And unless we think along the lines of Hartmann, i.e. that with increasing intellectual and cultural development reason will triumph over Will enabling the negation of existence, the appreciation of the tragic will be limited to the few. It will be recalled that we found considerable merit in Hartmann's belief that general 'progress' will also facilitate the advance of rejectionism. In any case, Zapffe's notion of the 'objective' nature of the tragic, i.e. the craving for a metaphysically meaningful and just world and the impossibility of its attainment, remains and will remain a fundamental feature, a void at the heart of human existence. For Zapffe living with this truth is only possible through elaborate subterfuges, in short lying and self-deception. The alternative is to acknowledge the insoluble metaphysical and moral problems of human existence and to bring this misadventure to an end voluntarily, once-and -for-all.

Although Zapffe's emphasis is on the metaphysical, i.e. the question of meaning, for him ethical awareness ('the brotherhood of suffering') is also a part of the problem of human consciousness. To live is to suffer and humans not only have to contend with their own suffering but also with the awareness of the pain and suffering of all other creatures. Homo sapiens will do no harm and do themselves and the environment a lot of good (Zapffe was an early ecologist who decried the destruction of nature with the advance of industrialization) by voluntarily disappearing from the earth. As he wrote, even a 'two-child policy could make our discontinuance a pain-free one' (quoted in Ligotti 2010, 29). What he found particularly objectionable is the doctrine that 'the individual "has a duty" to suffer nameless agony and a terrible death if this saves or benefits the rest of the group'. For Zapffe 'no future triumph or metamorphosis can justify the pitiful blighting of a human being against his will' (29).

Zapffe practiced what he preached remaining childless on principle. Unlike earlier rejectionists, including Schopenhauer and Hartmann, Zapffe's route to liberation is by way of non-procreation. This makes him a 'modern' rejectionist, i.e. secular, rational and with a preventative approach. As we shall see below this approach to existence has been developed and argued in detail by David Benatar.

David Benatar: Philosophy of Anti-Natalism

Unlike Schopenhauer and Hartmann, David Benatar, a contemporary philosopher, is not a metaphysician. He is not concerned with questions such as what is the fundamental nature of reality (Schopenhauer) or whether there is a teleological principle at work in the world leading towards a goal such as the negation of existence (Hartmann). What he shares with these two 19th century philosophers is his interest in exploring, indeed his passionate engagement with, the question of existence which he finds problematic in that it invariably entails a great deal of pain and suffering. In common with these earlier thinkers he too seeks a way out of the pain and suffering of existence (Benatar 2004 and 2006).

Although Zapffe was also an anti-natalist, Benatar is unique in his focus on procreation and in his strong advocacy of anti-natalism on philosophical grounds. He holds procreation to be an immoral act in that it inflicts gratuitous pain and suffering on someone who has not asked to be born and who is brought to being primarily to serve the interests of others, including of course the parents. While other philosophers have touched upon these issues, Benatar's work is the first comprehensive and detailed treatment of issues surrounding procreation or what he calls 'creating people'. It is a work of modern philosophy in that, written in the 21st century, it is free of metaphysical assumptions and relies entirely on reasoning and empirical evidence for its arguments. On the other hand it is in line with traditional philosophizing in that it does not shy away from value judgment with regard to the nature of existence in general and human existence in particular. Benatar uses the term 'analytic existentialism' to characterize his work, a term that encapsulates both his method and the nature of the problem he tackles (Benatar 2004, 1-3). In short his work uses the methodology of 20th century English philosophy in order to grapple with issues of existence.

The term 'existentialism' has been historically associated with Continental philosophers of the 19th and 20th centuries, such as Kierkegaard, Heidegger and Sartre. It is they who have been concerned with issues of human existence. However their approach has often been rather "expressionist" and rhetorical, insightful but not characterized by analytical rigor and logic (1-3). Coming from a philosopher from the English-speaking world Benatar's book <u>Better Never To Have Been (2006)</u> is in this regard a path-breaking work. Furthermore as Benatar points out not only English philosophy of the 20th century but also Continental philosophy of existentialism has had little to say about procreation and the ethics of bringing new people into the world (9). Much of that philosophy has been concerned with those who already exist, rather than with the philosophical issues raised by the creation of new lives. An important difference between the attitudes and the approach of 19th century philosophers, such as Schopenhauer and Hartmann, and 20th century existentialists such as Heidegger and Sartre, has been the latter's emphasis on the individual and his manner of being (see e.g. Sartre 1948; Watts 2001, 55-6).The burden of making sense of existence and choosing to act in a particular way is laid squarely on the individual. The only value upheld by the existentialists is that of authenticity, i.e. that individuals choose their life course and course of action in the fullest awareness of the situation and the consequences of their choice. To simply follow conventional norms, religious injunction or any other external authority is to act in 'bad faith' (Sartre), or to be 'inauthentic' (Heidegger). Since existence precedes essence (Sartre), the creation or affirmation of values and making sense of existence becomes an individual act. There is no escape from this 'dizzying' freedom . The existentialist philosopher steers clear of value judgment, at least in any explicit manner, concerning life and its significance in general.

By contrast Benatar's approach is one where he commits himself, he takes a stand. He lays his cards on the table and is quite open about his assessment of the nature and value of existence and seeks to convince others to act in accordance with his perspective and beliefs. His conclusion is that procreation is an immoral act which brings significant harm to lives which could and should be spared that harm. It is time, however, to move beyond these introductory remarks to a detailed look at his substantive thinking about existence.

The Asymmetry of Pain and Pleasure: Unlike Schopenhauer and Hartmann Benatar starts not with the big picture, e.g. the nature of the world, but what might be called the micro-philosophy of procreation. His starting point is that all sentient beings suffer some harm or 'bad' in their lives and many suffer significant harm. However it is only humans who have the consciousness and the ability to *prevent* this harm by not creating new lives. For no matter how lucky a life might be it is bound to undergo some pain and suffering. True, lives also consist of pleasures or 'goods of various kinds. But whereas we have a moral obligation not to inflict harm on anyone if we can help it we have no corresponding obligation to bestow pleasure on future people. Thus by refraining from procreation we prevent harm to future people but because of the asymmetry between pain and pleasure we do nothing immoral in depriving such people of the pleasure they may have had had they been brought into being. If conferring pleasure on future people were to be a moral obligation we would have to have as many children as possible. This asymmetry of procreational morality is at the heart of Benatar's philosophy (Benatar 2006, 28-31). In support of this fundamental asymmetry he refers to four other asymmetries considered as valid and normal by people. It is of course related to his view of existence as a source of pointless pain and suffering inflicted on all sentient beings.

Largely implicit in Benatar's view of life is the notion that whatever we consider as the positive aspects of life they do not in any way justify the 'cost' in terms of the inordinate amount of pain and suffering involved. However his concern is not with existing people and their emancipation from the shackle of existence in the manner of the Buddhist nirvana or Schopenhauerian renunciation of the will. It is *prevention* rather than cure that his thinking is aimed at. As we shall see later he is not opposed to suicide, especially rational suicide. But his centre of attention is procreation – the means through which people are brought into being and subjected to the unnecessary pain and suffering of existence. Indeed not only the progeny suffers but in so far as it becomes itself a source of further proliferation of lives each new life represents 'the tip of a generational iceberg of suffering' (6). For example, if each couple begets three children, in ten generations that mounts up to over 88,000 people which 'constitutes a lot of pointless suffering' (6).

The Immorality of Procreation: A child may simply be conceived as a byproduct of copulation. Here it is simply *coital interest* which gives rise accidentally, as it were, to a child. With the development and widespread use of contraception coital interests can be satisfied without resulting in procreation. But even where a child is conceived intentionally it is not usually for the sake of the future child itself. Rather the guiding motive for having children is the parents' own interest. This can take many forms. One may wish to have one's own genetic offspring for the sake of biological reproduction and continuity of oneself, i.e. passing on one's genes to the next generation. It may be out of parenting interest, i.e. to have the experience of nurturing and raising a child and establishing a lifelong bond with the child one has raised. When grown up the child can act as a support for the parents in their old age. One's property, title, social status and the like can be passed on to the child. *Parenting interest* is different from *reproductive interest* in that the former can be satisfied through adoption although most people prefer to do it through reproduction. Apart from these direct interests and motives for procreation there may be other considerations that favor natalism. These may be religious – raising a family as a duty, economic, e.g. state policies aimed at increasing the working age population, political and cultural, e.g. in the interest of preserving or increasing the size of a nation or tribe. This is only a short list of reasons for having children. Many others could be added.

The main point is that children generally serve as a means to various ends whether the parents' or others' (96-8). They are not brought into being for their own sake, i.e. we do not confer life on someone simply for his or her own good. And if we believe that we are doing a favor to someone by bringing them to life then, argues Benatar, we are totally mistaken (97). For the pain and suffering that is in store – it may be more or less – for that child is not worth the 'pleasure' or other goods of life that might also come its way. Benatar's main point is that one who does not exist does not miss the 'good' things of life but one who is born is sure of being exposed to the evils of life. And if we leave aside various extrinsic interests, including those of the parents, in procreation, i.e. which treat the child as a means to other ends, then there is no case for bringing a new life into the world. Indeed given that to bring any being to existence is to expose it to at least some degree of harm, and often a great deal of harm, it becomes a moral duty to prevent this harm if we can. Hence refraining from procreation becomes a moral duty and by the same token procreation becomes an immoral act. To this the rejoinder could be that if the parents feel happy about their own lives and are glad to have existed then it is a reasonable assumption that their child would also feel the same. But this does not necessarily follow, argues Benatar. We may feel glad to have existed at this moment but feel otherwise at another time as we go through our lives, grow older or face difficult times. Thus our opinion and attitude to life at any given moment is not a reliable basis to make that judgment. On the other hand, we do not know in advance the child's own preference. But should we not give it the benefit of the doubt? And Benatar suggests using the famous 'maximin' principle of John Rawls (1971) in this situation (Benatar 2006, 178-82).

Maximin is concerned with arriving at what might be considered a just social order through a hypothetical construct. The basic idea is that people come together in order to devise a fair system of distribution of life's resources in the 'original' position, i.e. under a 'veil of ignorance', so that they have no knowledge in advance as to how each will be positioned in real life, e.g. whether they will be born rich or poor, intelligent or otherwise. This ensures impartiality in decision making since no one knows in advance what fate has in store for them. In this situation, argues Rawls, rational individuals would seek to maximize the minimum, i.e. the resources available to the most disadvantaged persons. In short, each would act as if he or she might be the one born disadvantaged and so ensure that certain minimum conditions of life are made available to such individuals.

If we apply the maximin principle to the choice of existence for the unborn the rational choice, argues Benatar, would be to choose non-existence. Since no one knows in advance how much pain and suffering will come their way they must proceed on the assumption that they might be the one exposed to most suffering. The only way to ensure that one does not suffer that fate is to choose not to be born. Some people have raised the question whether these hypothetical individuals should at least know the probability of being one of the worst off so that they could then decide with fuller knowledge of the situation they face. However in the application of his principle of maximin Rawls explicitly forbids this in the interest of strict impartiality. In any case, argues Benatar, even the knowledge of probability does not affect his argument since every life must face *some* measure of harm and it is only a question of facing more or less. Hence concludes Benatar, 'it is always irrational to prefer to come into existence. Rational impartial parties would choose not to exist'(182).

The Moral Obligation of Non-procreation and the Right of Reproduction: Clearly the moral obligation of not bringing new people into existence goes against the conventional wisdom that having children and 'founding a family' is a good thing. It also goes against what has been recognized in the UN Charter as a basic human right (102n). How can a fundamental human right be immoral? Can

Benatar's viewpoint be reconciled with the right of reproduction recognized by the UN Charter of Rights? Benatar believes that it can be. For what the UN Charter does is to proclaim the *right* to reproduce. It does not preclude choice, i.e. the right *not* to reproduce. However the fact that something is legal does not preclude it from being immoral. For example in South Africa racial discrimination was enshrined in law but it was clearly an immoral act. During the heyday of slavery the institution was legal and affirmed the right of slave owners over their chattel. Clearly it was an immoral institution. Historically there are many instances of forms of behavior being considered immoral and/or illegal which were later deemed not to be so, e.g. homosexuality. The formal right of procreation could be enshrined in law but the act of procreation could be deemed immoral from the viewpoint of existence that Benatar holds (102-3, 111), . Benatar makes it clear that he is not advocating a state ban on reproduction or any such intrusion into the freedom of citizens and state suppression of rights. But he insists on the *moral* duty not to procreate as a voluntary act on the part of would be procreators. It is in the nature of a moral commandment, viz. that 'thou shall not inflict pain and suffering on a sentient being by bringing it into existence'.

Seen somewhat differently there might be a conflict of rights involved here. Against the so-called reproductive rights of adults we need to consider the right of the putative individual, i.e. the right of the unborn, not to be brought into existence. A common basis for denying such a right is the argument that 'prior to procreation that person does not exist and thus there can be no bearer of the right not to be created' (53). Benatar thinks this may be an unduly narrow view of rights. For if one could be harmed by being brought into existence, then there could be a right of protection against such harm even if it is a 'right that has a bearer only in the breach' (53). Benatar's focus is on the *duty* not to procreate but he believes there could be a case for recognizing the *right* of the unborn to be spared procreation. The application of the principle of maximin to elucidate the preference of the unborn (see above) can be seen as an indirect recognition of such a right. But he does not follow this line of argument any further.

In sum, the interest of the parents and others who favor procreation appears to be in conflict with the interest of the putative child in not being brought into existence. True, for humans the desire to reproduce, implanted by nature, is presumably a strong one and reproductive right affirms this basic human trait. However it ignores the other side of this right, viz. the putative violation of the right and the resulting plight of the child which follows from the satisfaction of this urge.

<u>Benatar's View of Existence and the Pollyanna Principle:</u> Benatar's view of existence as something that inflicts serious harm on sentient beings is clearly in conflict with the positive view of existence that the vast majority of people hold. A part, if not a good part, of this discrepancy can be explained by what Benatar calls the Pollyanna principle. Put simply it is a bias towards optimism, a tendency to put a positive spin on life and one's experiences. In Benatar's words 'If coming into existence is as great a harm as I have suggested and if that is a heavy psychological burden to bear then it is quite possible that we could be engaged in a mass self-deception about how

wonderful things are for us' (100). In other words we resort to a form of false consciousness about the nature of existence.

A number of factors seem to be at work here – some biological and others socio-cultural. Optimism is in line with evolutionary success and survival. Thus 'hope springs eternal in human breast'. Pessimism, on the other hand, is likely to result in a tendency to withdraw from the struggle for existence, even an inclination towards suicide or at least a refusal to procreate. Conventional wisdom instills in us the virtues of a positive attitude towards life. Homilies such as 'look at the bright side of things', 'get on with it', 'no use complaining', 'be thankful for your blessings', are a small, if rather crude, sample of the repertoire of statements in this vein. And indeed if one is brought into existence to play the game of life one needs every encouragement and motivation to take the game seriously and to try to play it well according to the rules.

The Pollyanna principle shows itself at work in a variety of ways, preeminently in the positive self-assessment of one's quality of life. Thus when asked to recall events from their lives people recall far greater number of positive than negative experiences (65). This reflects how we distort the judgment of how well our life has gone[5]. Projections about the future also tend to exaggerate how good things would be. Similarly self-assessments of current well-being also show a marked positive bias. Thus an overwhelming majority of people claim to be 'pretty happy' or 'very happy' (66). Within any given country the poor are almost as happy as the rich. Benatar points out that a well-known psychological phenomenon that contributes to the positive bias is what can be called accommodation, adaptation or habituation (cf. Zapffe above). Thus if our condition gets worse we express dissatisfaction. But with time we tend to adapt to the situation and lower our expectations. Finally there is an important aspect of self-assessment of well-being that can go unnoticed. It is the comparative or relative nature of our judgment. It is not about how well things are in themselves or how well they have gone with oneself. It is rather how well they are in comparison with others[6]. One of the implications is that those negative features that are shared by everybody else may be ignored in our self-assessment. These include the frustrations, inconveniences and disappointments experienced in everyday living (72).

<u>Assessment of the Quality of Life:</u> For a more systematic approach to the assessment of quality of life Benatar examines three types of theories concerned with the issue: Hedonistic theories which focus on the balance of pain and pleasure in a life, or more precisely negative and positive mental states associated with these; Desire-fulfillment theories which consider the extent to which our desires are fulfilled; thirdly, Objective list theories which judge lives in terms of the presence of good and bad things, things that these theories consider significant irrespective of whether they bring pleasure or pain. Benatar draws our attention to various shortcomings of these theories and the resulting assessment of the quality of life and argues that, in any case, none of these succeed in convincing us that existence does not involve significant harm (69-70).

For example with regard to the Objective list theories Benatar finds that they are constructed from a relativistic or humanitarian perspective rather than *sub specie aeternitatis*. Thus they are more useful in comparing one life with another, e.g. in terms of creativity, freedom, deep personal relations, having children etc. What the theory does not tell us is how good human life per se is. Objective list theories of course differ with regard to the items they include in their list as of value for individuals to possess or enjoy. An important item missing from these lists is that of a meaningful life. The desire that life should have a 'meaning' going beyond simply existence and reproduction seems essential for most human beings. But looking at life *sub specie aeternitatis* it is clear that conscious life which is 'a blip on the radar of cosmic time is laden with suffering – suffering that is directed to no end other than its own perpetuation' (83). Seen in this wider perspective life has no meaning.

We should note that Benatar does not mention religious belief in this context. Undoubtedly that is one of the sources of meaning ('anchoring' according to Zapffe) for many people and is likely to be one of the values on some objective list theories. The omission of religion seems to suggest that Benatar considers it as a form of false consciousness or bad faith rather than an objective basis which could provide meaning *sub specie aeternitatis*. He notes that people try to find a meaning to their lives from a humanist perspective, e.g. service to others, realization of some personal goal, creative endeavor and the like. But it would be much better, argues Benatar, if life in general had some meaning independently of a human perspective, i.e. if it mattered from a transcendental perspective. Since there is no such meaning and yet human beings yearn for such a meaning this void becomes a perennial source of anguish of conscious existence adding to other forms of suffering (82-6). The reader will note here the similarity to the main issue raised by Zapffe (see above).

Benatar comes to the conclusion that none of the three theories concerned with quality of life can make a case for existence being superior to non-existence. While they may be useful, up to a point, in *comparing* the quality of individual lives with one another they do not address the problem of existence from an absolute or universal, as distinct from a relativistic and humanitarian, perspective. The case for not inflicting the considerable amount of harm that every life must suffer remains strong.

Sufferings, Human and Animal: In further support of his view of existence Benatar provides a harrowing account of human suffering through the ages. These include natural disasters, famines, wars, diseases and epidemics as well as privately and publically inflicted cruelty, torture, rape, murder and other forms of killings. Thus according to one estimate, during 1900-1988, some 170 million (and possibly as many as 360 million) helpless citizens as well as foreigners were the victims of all kinds of brutalities and killings *perpetrated by their governments* !(91). The 20th century was the bloodiest on record in terms of wars. Conflict-related deaths numbered 110 million compared with 19 million in the 19th and 7million in the 18th century. There are of course many other forms of pain and suffering inflicted by humans individually on each other, ranging from assaults to murder (91).

If we add to human suffering what billions of animals go through, the suffering inflicted on them by humans – for eating, experiments or other uses – and by other animals we see the vast world of suffering that existence involves. It is a condition we choose to ignore and desensitize ourselves to. Optimists and 'cheerful procreators' try to put a positive gloss on the human situation. But in view of the 'amount of unequivocal suffering the world contains' they appear to be on very weak ground (89).

<u>Suicide and existence:</u> _Benatar's line of thinking on suicide is altogether different from those of Schopenhauer and Hartmann (see Ch.2 above). Although in common with these philosophers he too is against suicide, as a way out of existence, his reasoning is very different. But let us start on a personal note. Some of Benatar's critics have made an ad hominem attack on him arguing that if things are really as bad as he thinks they are why is he still with us? Why not 'put an end to things now?' (Belshaw 2007). Indirectly then Benatar stands accused of hypocrisy in spite of his logical and quite reasonable objection to suicide as a solution. Indeed Schopenhauer also faced the accusation, directly or indirectly, that his own practice belied his preaching. While his ideal for salvation was the abnegation of the will and asceticism his own will to live remained strong, he craved recognition and he was no ascetic[7]. He was known to have enjoyed dining well regularly at a good local inn.

Benatar, it is true, believes existence to be a serious harm. But he does not believe suicide to be the solution. His focus of attention is on *future lives* rather than *present* ones. Indeed an important feature of Benatar's philosophy is the distinction between *starting* new lives and *continuing* present lives. For him the road to liberation from existence is anti-natalism: to eschew procreation and spare new lives the pain and suffering that they must undergo. But those who exist already are in a quite different situation. Growing into adulthood they have developed strong interests in continuing to exist. Harms or conditions that make life not worth continuing must be sufficiently severe to defeat those interests. These include personal, emotional and social relationships which involve family and friends. Suicide will cause a great deal of pain and suffering for the bereaved. To quote Benatar, it can have a 'profound negative impact on the lives of those close to one' (220). And although the deceased himself is beyond the reach of pain and suffering the bereaved suffer harm. Thus existence is a form of trap. It shows how glib it is to argue that those who are not pleased with existence can simply put an end to their own life. Clearly it is not as simple as that. For quite apart from other things, once a life is started an irrational attachment to existence is implanted within it thus erecting a major obstacle to suicide. As an old lady, a character in Voltaire's *Candide*, expresses it, ' A hundred times I wished to kill myself, but my love of life persisted. This ridiculous weakness is perhaps one of the most fatal of our faults. For what could be more stupid than to go on carrying a burden that we always long to lay down? To loathe, yet cling to existence?' (quoted in Benatar 2006, 219-20).

Suicide raises another important issue. It differs from procreation in a crucial respect. The latter involves making decisions for *others*, the unborn, beings who cannot make the judgment whether to come into existence or not for themselves. Suicide, on the other hand, involves a decision made by an adult about his *own* life. And for Benatar, there can be no objection in principle to an adult, responsible for his own acts, deciding to end his life. Nonetheless the web of human relationships, consideration for others, and many other vested interests make suicide problematic from a practical as well as a moral standpoint. No such problem arises with abstaining from procreation. And the pain caused by childlessness to oneself and others is undoubtedly 'mild in comparison' to that caused by suicide (220). In other words, suicide is an attempt to seek a *cure* to a condition through a form of violence to oneself and others whereas Benatar advocates *prevention* which involves no violence. Not bringing new beings to life is a gentle and morally acceptable way to non-existence. Non-procreation is an act of altruism which *prevents* harm to others whereas suicide is an act of egoism which causes harm to others and to oneself. This is a crucial distinction.

Human extinction: the logic of anti-natalism: One of the main characteristics of Benatar's philosophy is its attempt to follow through the implications of his stance on existence. Thus the logic of his anti-natalism, i.e. the imperative of stopping procreation, is the gradual disappearance of humans from the world. Benatar is fully aware of this and considers how the cessation of human race could be managed to minimize the additional suffering involved in the process.

However, he starts from a different position. His premise – and he claims this is based on scientific grounds – is that the human race is sure to go extinct, sooner or later. It is not a question of *whether* it will but rather when and how (194-5). For Benatar the sooner this happens the better since it will save millions, if not billions, of human lives from suffering the harm of existence. Thus at the current (2006) rate of reproduction, a billion people would have been added to the number of Homo Sapiens in just twelve years (165). Immediate extinction could save a great deal of suffering. But this is unlikely, short of an astronomical accident, e.g. strike by an asteroid, or some other cataclysmic natural event. It is more likely to be a messy, drawn out process – as a result of our action such as a nuclear warfare or environmental degradation.

In Benatar's view a phased extinction, planned in a way so as to reduce suffering should be the more humanitarian and compassionate way. Planned and gradual extinction can take care of or at least reduce the adverse impact of non-procreation, e.g. the decline of younger population and a major shift in the ratio of the young to the old, producers to non-producers. Demographic changes will create problems of maintaining a functioning and viable society. If procreation were to stop altogether some additional people may have to be created than would otherwise be the case in order to sustain some measure of quality of life for the later generations as the process of human decline continues. Even in such a situation, however, there is no doubt that the last generation of humans will likely undergo a great deal of suffering (197). Although their situation will be a bleak one indeed, it is hard to know 'whether their suffering would be any greater than that of so many people within each generation' in the normal course of events (198). In order to determine whether the regrettable future of impending extinction is bad, all things considered, we have to take into account not only the final people's interests but also of the harm avoided by not producing new generations. It is undeniable that whenever humanity comes to an end there will be serious costs for the last people. But all things being equal nothing is gained if this happens later (198).

One of the arguments against the extinction of human species is that as a result much that is valuable and unique will be lost. For example morality, reasoning, diversity (Benatar does not mention the appreciation of beauty, creativity and other aspects of human culture and civilization) will disappear. But, asks Benatar, 'what is so special about a world that contains moral agents and rational deliberators?' (199). Humans attribute value to many things including the presence of beings such as themselves and their achievements from an 'inappropriate sense of self-importance' .If humans disappear then these things will also disappear and no one will be there to regret their absence. For seen *sub specie aeternitatis* these things do not appear to have any value. In any case 'it is highly implausible that their value outweighs the vast amount of suffering that comes with human life' (200). The concern about the non-existence of humans is 'either a symptom of human arrogance that our presence makes the world a better place or is some misplaced sentimentalism' (200). However Benatar is under no illusion that human species will take the path of voluntary extinction. He emphasizes that his approach is one of principle and theory. He is spelling out the implications of his viewpoint including what might be involved in its practice (184).

Benatar, an assessment: Let us begin with Benatar's view of existence. It is similar to those of Schopenhauer and Hartmann. The emphasis is on the pain and suffering entailed by existence, felt most keenly by sentient beings. For Benatar, too, the immense suffering that life inflicts on sentient beings can in no way be justified by its pleasures and other positives. However in one respect Benatar goes further. For him *any* suffering, however small, rules out bringing new lives to being. Since all beings suffer *some* harm by coming into existence starting new lives can never be justified. In other words since Benatar is mainly concerned with procreation, sparing the unborn the pain of existence is not a question of the calculus of pain and pleasure. The unborn cannot be deprived of any pleasure. On the other hand bringing a being to life is sure to inflict some harm on it. This 'asymmetry' between pain and pleasure is one of his major contributions to anti-natal thought.

Prevention of suffering through non-existence: If Benatar's view of existence is not dissimilar to those of Schopenhauer and Hartmann it differs quite sharply from their thinking about liberation from existence. This in turn is related to a fundamental difference between Benatar and his predecessors which underlines his modern approach to the question of existence and liberation. It is that his thought is free of metaphysical assumptions. It does not presuppose any fundamental reality behind empirical phenomena, e.g., the Will (Schopenhauer), the Unconscious (Hartmann), transmigration of souls and rebirth (Hinduism and Buddhism). It is also free of any mysticism or mystique, e.g. those associated with the notions of moksha, nirvana, and arguably also willlessness. What he advocates is liberation through non-procreation. This route or mode of emancipation is secular and realistic.

His anti-natalism is informed by compassion for and empathy with all living things, especially human beings, that are born to suffer and in this it resonates with Buddhism. His emphasis on life as gratuitous suffering and the imperative to liberate humans from it are also in harmony with the Buddhist approach. Where he differs is in his concern with *future lives* rather than *present lives* and hence also in the path to liberation that he advocates. It involves no asceticism and makes no demand for abstinence from sensual including sexual gratification. Given contraception sexual needs can be fulfilled without giving rise to reproduction. One can lead a 'normal' life in every way but one, i.e. not having children. His clear distinction between present lives, which do not need termination, and future lives which ought to be prevented from coming into existence and thus spared life's agony makes for a modern, secular and practical approach which, however, mirrors the older conception of moksha and nirvana. It is this modernity that is both Benatar's appeal as well as his major contribution to the principle and practice of the prevention of suffering to human beings. It may be seen as a form of modernization and democratization of the notions of moksha and nirvana. To refrain from procreation is within the reach of every individual, and compassionate concern for future lives is a morally worthy basis for such action. Since it is not self-centered but rather oriented towards others it is *altruistic* rather than *egoistic*. Except for the philosophies of Hartmann and Zapffe, the others examined above involve ego-centrism and elitism. Only a very small select group of virtuosi are capable of the asceticism - bordering on self-torture - and the dedication required for liberation. With Benatar the situation is very different and the 'sacrifice' demanded is far more modest.

Prospects of anti-natalism: However, despite the strength of his arguments and the compassion and altruism underlying his approach Benatar believes that his ideas will have little influence, and that 'baby-making' will go on as before (62). Although he does not spell out the reasons for his belief, clearly the assumption is that false consciousness regarding existence, the Pollyanna principle, the 'instinct' for survival and reproduction, social norms and their constant reinforcement will make the vast majority continue on the conventional path.

Although he is right in his broad assessment of the situation he appears to underestimate the *potential* for an increasing awareness of the harms of life and the acceptance of anti-natalism. A variety of factors are involved here. First, there is the greater ease of worldwide communication of ideas especially through the internet. Thus Benatar's book has received wide publicity and generated a good deal of discussion through this medium. Second, there is greater freedom of choice and awareness of existential choices available to us. Third, there is a declining taboo against voluntary childlessness. Benatar pays insufficient attention to the fact that for several decades now anti-natalism has been on the rise. Increasing number of women and men are deciding not to procreate. True, much of voluntary childlessness is motivated by factors other than philosophical. Often it is a life-style choice, the wish to remain 'child-free' in order to follow wholeheartedly some personal objective or goal or simply on account of the feeling that one is not interested in being a parent. For many women voluntary childlessness is a road to freedom. It involves rejecting the historic female role of a mother and wife, and having the opportunity that men have always had to do other things. Despite social pressures, despite government attempts in the developed countries - where voluntary childlessness has made rapid advance - to provide financial and other inducements it is unlikely that the tide of anti-natalism can be turned back.

Ideas, technology and economic development are all playing their part in promoting anti-natal attitudes and choices. In particular we should mention secularization, the decline of religious control of reproductive freedom, advances in contraception, and finally economic and educational development in enhancing women's opportunities for employment and the pursuit of a career. We may reasonably assume that philosophical anti-natalism is also making progress though unfortunately we have no idea of the number of individuals or couples who have chosen not to reproduce on philosophical grounds. As far as one can surmise philosophical childlessness is likely to be a very small percentage of voluntary childlessness which itself remains a small proportion of natalism and involuntary childlessness taken together[8]. However we must not think of this as a static situation but rather as an evolving state of affairs. In sum, Benatar may be right to be somewhat 'pessimistic' in the short-run regarding the influence of his ideas and the spread of anti-natalism. But in the long run the prospects are definitely better, or at least more promising.

It is important to remember that throughout the ages only a small minority of people have refrained from procreation, and the acknowledged vehicle of such rejection has been religious, usually taking the form of a celibate priesthood or monkhood. However in early Christianity, early Buddhism and in Hinduism more generally, it took the form of wandering monks or holy men. But we know very little about *lay people*, whether in India or in the West, who have on the quiet followed a similar path on philosophical grounds. It would have meant remaining unmarried and childless while in other ways leading a regular life. Benatar's book, for example, brought forth responses from people who have come to conclusions similar to his and have refrained from procreation without publicizing their principles and practice[9]. Indeed we would argue that non-procreation would be far more practicable and likely today, in the context of the social change we outlined above, than in the past. Again lacking any studies of the phenomenon of philosophical childlessness we have virtually no idea of the number of people involved, whether at present or in the past. True, we know of many intellectuals and philosophers who have in principle rejected procreation. But they too generally prefer to remain silent about their choice. The few exceptions we come across, e.g. Flaubert (quoted in Benatar 2006, 93) and Cioran (see above), seem to prove the rule. What we can say, however, is that refraining from procreation, often linked to the rejection of existence itself, has a long history involving both religious and lay people. We have chosen to call this the philosophy of Rejectionism. True, this view of life has been restricted to a small minority of people but, we believe, it is a minority likely to grow in coming years.

In this context we need to acknowledge the prescience of Hartmann's thought. The relevance of his ideas for Benatar's anti-natalism should not be ignored. He may have been utopian in looking for a world-wide move to reject existence but his belief that progress, which includes the spread of education, communications, and secularization, will lead to the growing strength of reason over will and undermine the latter's grip on life has a good deal of credibility.

<u>Anti-natalism as a Social Movement:</u> Of course Benatar is not concerned with the sociology of anti-natalism or its advance as a

social movement. His focus is almost entirely on the *philosophy* of anti-natalism, i.e. with reasoning and evidence in support of his arguments against procreation. We should note however that he does pay some attention to issues concerning the practice of anti-natalism. For example, he writes that 'foregoing procreation is a *burden* (italics added) – that it is a lot to require of people, given their nature' (101). His choice of the word 'burden' suggests that non-procreation does mean hardship and deprivation, including psychological, for the individuals concerned. Although he does not spell out the nature of the burden, his earlier remarks (98) indicate that children are created to serve a variety of significant interests of the would-be procreators. Indeed even those who accept the argument that existence entails significant harm, not redeemed by anything that is positive about existence, may find it difficult to lead a childless life given that children serve a wide variety of functions for the parents. True, thanks to contraception and legalization of abortion coital interests can now be satisfied without involving procreation. With a partner who accepts one's view of life it is also possible to lead a married life. Adoption too is available as a substitute to procreation.

However in so far as the childless must rely on other people for practically everything they may also feel that they are being 'free riders' (see Smilansky 1995, 46). They are enjoying the benefit of having other people's children serve their needs. Others have, so to speak, done their 'duty' to reproduce and replace the population but they have not. They are benefitting from others' sacrifice and hard work of reproducing and parenting - although the latter have also reaped the reward and satisfaction that comes from raising a family.

Quite apart from this, there is also the deprivation and isolation that the childless may feel in old age when, perhaps also disabled and ill, they cannot turn to their children for help and support. True, they may have siblings, nephews and nieces and other relations or friends to provide them with care. Moreover studies show a somewhat mixed picture in respect of aging and childlessness. Having children does not guarantee care of the aged nor is the converse true[10.] Moreover the childless, as indeed others, may quite legitimately look to the welfare state and voluntary organizations for personal care. And in this regard the situation of the childless is far better today than in the past. Nonetheless it is still the case that the family is expected to play an important part in the care of the ill and frail elderly albeit in partnership with the state. Here again the childless may feel being a 'free rider'. The point is that once you come into being you are enmeshed in interdependencies and are subject to social norms that are difficult to ignore.

Admittedly one can easily exaggerate these problems. Clearly the childless make their contribution to society. They work, they may be more active in the community, they pay more in taxes not having the benefit of child tax credits and the like, they do not burden the state with the health and education costs involved in child birth, child health care and the costs of schooling and post-secondary education of children. Furthermore, they may be more inclined to leave a legacy to philanthropic and other charitable organizations which benefit society at large. So clearly there is a tradeoff involved and there is scarcely any objective basis to feel guilty about voluntary childlessness. The childless have every right to have their needs met. Moreover they also have the satisfaction of acting unselfishly, i.e. not inflicting life and suffering on another in order to serve one's own interests. And seen from a global perspective not adding to the world's population is an act of altruism

However there are other problems faced by non-procreators. True, if you are deeply convinced of the evil of existence you may be prepared to face the difficulties – some of which we noted above - entailed by your decision. But consider some other issues raised by the practice of non-procreation. The average age at which a young couple may typically 'fall in love' and decide to get married is likely to be in the twenties or early thirties. Youth, by its very nature (physical instincts and life force) is full of hope and the promise of happiness. Sexual attraction and the urge to reproduce is strong. Having a child may be a part of the sexual union and bond between the partners. As Schopenhauer noted, romantic love is a ruse of nature to make people reproduce and continue the species. Many fall victim to this ruse. Can the appreciation of the evil of existence, with all its pain and suffering, come easily at an early age? Again we know very little about this. Anecdotal evidence suggests that lifestyle-based anti-natalism can come at a young age. And in this regard there is evidence of a generational shift. The young generation is far more likely to eschew procreation. On the other hand philosophical anti-natalism, If it comes at all, is likely to be a product of a more mature age. And by then for many the deed is already done. Enlightenment may come too late to forestall procreation[11].

Benatar is not unaware of the many difficulties that lie in the way of the progress of philosophical anti-natalism, both as belief and as practice. His 'pessimism' regarding the scant influence of his ideas shows this clearly. All the same we must note that his accent is on the beliefs and ideologies that are available to humans to justify continuing 'baby making' rather than on the problems facing the rejectionists. As regards the former, put simply it is a blend of the natural will to live, parental interests, socialization, and the Pollyanna principle. Benatar's notion of people being in a state of false consciousness about life being a good thing - in the face of much evidence to the contrary - is a result of these beliefs and ideologies. But how to counteract these ideas and norms? There are a number of issues here. First, how many people are likely to be aware of the philosophy of anti-natalism, e.g. through reading Schopenhauer, Benatar or other rejectionists and visiting the relevant websites? How many reading lists, undergraduate or graduate for that matter, in the universities would feature Benatar's book? Judging by the critical reviews of his book, and especially his key arguments, it appears that most academic philosophers are unsympathetic if not hostile to Benatar's philosophy. Second and more generally, how many people would be convinced by the arguments and evidence advanced in these books and other sources? And as pointed out above, by the time people are 'ready' for the message it may be too late as far as they themselves are concerned. However the widespread publicity and attention received by Benatar's book suggests that it has played an important part as a catalyst in stimulating awareness and debate, and many more people may be ready to accept rejectionist beliefs. Finally there is the problem of putting such beliefs into practice, a problem to which Benatar is unable to pay much attention.

Admittedly it is quite unrealistic for us to expect Benatar to treat every aspect of anti-natalism, including the propagation of the message of non-procreation. Yet the practice of his philosophy requires further thought as well as research besides the compulsion of arguments and logic. Benatar has laid the foundations of philosophical anti-natalism, i.e. modern rejectionism, and it is up to others to build on it and contribute further to the principles and practice of the same.[12] We shall return to this question later (see Ch.5).

Endnote – Chapter 3

1. On existentialism see Introduction.

2. See e.g. Sartre (1948), Camus (1975). On Heidegger see Watts (2001). On Unamuno, Camus and Cioran see Dienstag (2006). See also Ligotti (2010, 47-50).

3. On the difficulty of getting a grasp on Cioran's thought see Kluback and Finkenthal (1997, 1-2, 11).

4. Although Schopenhauer is often described as an 'anti-natalist' we should note that what he espoused as the path to liberation was not refraining from procreation but willlessness, i.e. the abnegation of the will-to-live, and his concept of willlessness resembles the Buddhist notion of nirvana. Although Schopenhauer is explicit in his condemnation of procreation, which he sees as the strongest affirmation of the will-to-live, he does not advocate anti-natalism as the way out of existence. On this point see also Ligotti (2010, 30).

5. Hartmann makes a similar argument regarding our positive bias.

6. It is interesting to note Schopenhauer's observation on our positive self-assessment. He believes it is, in part, because we do not want to be an object of *schadenfreude*.

7. See e.g. Kierkegaard on Schopenhauer (http://philosophy.livejournal.com/1965165.html) downloaded 8/30/2011. See also Cartwright (2010, 534).

8. See Basten (2009).

9. This is referred to in Benatar's interview on Radio Direko (Radio 702/Cape Talk), 25 February 2009. See http://web.uct.ac.za/depts/philosophy/staff_benatar_betternevertohavebeen.ht

m accessed on 4/24/2013.

10. See e.g. Basten (2009, 12-13); Echo Chang et al. (2010); P. Span 'Aging Without Children',
http://newoldage.blogs.nytimes.com/2011/03/25/aging-without-children/ accessed on 6/5/2013.

11. A recent example is that of Jim Crawford (see note 12) who had two children before he became an anti-natalist.

12. Ligotti (2010) and Crawford (2010) are two recent contributions to the literature of rejectionism. Ligotti's work is a wide ranging and critical survey of the relevant literature from an anti-natal viewpoint. Crawford's book is an interesting combination of autobiography and the defence of anti-natalism. Neither of these writers, however, explores the issues raised by the *practice* of rejectionism, including its dissemination as a belief system. Very little seems to have been written so far on this subject. There is of course a vast and burgeoning literature on childlessness, especially as a choice, from the viewpoint of women's role and social identity. It does touch on the problems of practice. A recent work by Overall (2012), though not anti-natalist, is a comprehensive discussion of the philosophy - primarily the ethics - of procreation. It is written in a language that makes her book fully accessible to the non-specialist reader.

Chapter 4

Literary Perspectives

In this chapter we explore 20th century literary perspectives on the rejection of existence. The rationale for including literary perspectives alongside religious and philosophical ones was outlined in the Introduction and will not be repeated. Put simply, it is an exploration of the rejectionist perspective in modern literature. But why choose Beckett and Sartre?

Samuel Beckett is perhaps the outstanding writer of the 20th century whose work is, explicitly or implicitly, concerned with virtually all the major rejectionist themes encountered in the chapters above. The gratuitous pain and suffering that existence inevitably brings in its train, the intrinsic meaninglessness and pointlessness of existence, birth as the gateway to suffering and death, love and sex as traps for the prolongation of existence and suffering, the 'incurable optimism' of human beings in the face of the misery of existence, all this and more are to be found in Beckett's writing. Beckett's oeuvre – plays, novels, short prose pieces, essays – is fairly consistent in terms of its rejectionist attitude although there is a great deal of variation in both form and content (see e.g. Robinson 1969, Hamilton & Hamilton 1976). The similarity between Beckett's view of life and those of Schopenhauer and ancient Buddhism has often been noted (Buttner 2010; Bloom 2010, 3). Be that as it may, Beckett remains the 20th century writer with the most openly and thoroughly articulated rejectionist viewpoint.

With Sartre we are in a very different situation. He was not only a novelist and playwright but also a philosopher and a political activist. Moreover his world-view changed a good deal in the course of his life. From a somewhat anarcho-solipsistic writer in the 1930s he morphed into an existentialist philosopher in the 1940s, claiming that existentialism was a humanist doctrine which emphasized individual freedom to choose and to act in accordance with this choice. Later he became a left-wing activist and a communist fellow-traveler and later still, in the 1970s, almost a Maoist revolutionary. To some extent his writings reflect these changes. However his first published novel Nausea (1938), is a remarkable work on the rejection of existence, an encounter of human consciousness with material existence which the former finds, to put simply, quite intolerable. Nausea is the experience of feeling existence - one's own and others' - as a sheer contingent presence without any rhyme or reason. Its principal theme from our viewpoint is the absurdity and superfluity of existence but it has other rejectionist themes as well such as suffering and boredom, as well as the various justifications for existence that human beings employ. We begin, however, with Beckett.

Samuel Beckett: Literature of Rejection

Existence and Suffering: The basic theme underlying Beckett's perception of the world is the pointlessness and futility of existence.

Because of man's consciousness, his thinking 'self' – the 'I' – is aware of being tied to and being subject to the determinism of the laws of nature. Birth, maturity and eventual death, the compulsion of sexuality, the ravages of time – all these are aspects of existence that our consciousness makes us aware of yet over which we have no control. Moreover existence subjects us to all forms of suffering. Is there any point to it all? Why should human beings be thrown into the world and be dragged through this process – the business of 'living' which seems to have no purpose other than its own perpetuation through time? In this context birth and death seem equally 'meaningless' events. Some of Beckett's characters, e.g. in his novel <u>Watt</u> or his play <u>Waiting for Godot</u> (WFG), look for some significance or meaning to human existence but their quest comes to nothing. Beckett pours scorn over the idea that a Christian God or religion could confer meaning and reveal the mystery of existence. Nonetheless Biblical and other Christian allusions– myths, beliefs, symbols – occur frequently in Beckett's work and the counterpart to the notion of contingency is the absence of God.[2]

WFG, by far Beckett's best known work, which brought him instant recognition presents many of his concerns and themes. A tragicomedy, it is a poignant expression of man's anguish in conditions of meaningless existence. The play has little by way of action. It is primarily a dialogue between two tramps – Vladimir and Estragon – who are waiting on a country road for a character called Godot. Who is Godot or what he represents is left unclear. But the tramps are hoping for something like 'salvation', something that will confer meaning or significance to their existence. However Godot fails to arrive on the first day of their wait but sends word that he will surely come tomorrow. But the same thing happens the next day. Once again Godot fails to appear and sends word that he will definitely come the next day. That is where the play ends implying that the waiting will go on, as it has gone on in the past, although Godot will never come. In short man's quest for meaning, for finding an answer to the riddle of existence has gone on through the ages and will go on in the future. The hoping and waiting is as certain as its futility.

However while the tramps wait for Godot time will have to be passed and the boredom of existence contended with. Much of the play is about this. But they have done away with society and its usual trappings which provide most people with a 'meaningful' existence. (see especially Zapffe & also Benatar above). Having seen through the 'game' and refusing to play it Vladimir and Estragon are left on their own resources. They use language as a game to pass the time. They tell each other stories, pick quarrels, hurl abuses at each other. At times they think of suicide but do not take that way out and prefer to wait for Godot.

While waiting, the only "event" in the play is their meeting with a landowner called Pozzo and his menial, ironically named Lucky, who pass by. Pozzo decides to stop for a while and chat with the tramps. Pozzo's treatment of Lucky is cruel and humiliating to the extreme. Lucky is on a leash and is carrying a basket of provisions for Pozzo who wields a whip. Apart from other things Pozzo and Lucky illustrate the theme of man's cruelty to man. They pass by once more later in the play showing the ravages of time: Pozzo has gone blind and Lucky dumb. This time the tramps mistreat the blind and helpless Pozzo.

The tramps' decision to go on waiting, trying to seek an answer to the riddle, is very much in line with Albert Camus' injunction that man cannot but go on questioning the meaning of existence in spite of the silence of the universe. At least in WFG the tramps endure the agony and boredom of an apparently meaningless existence in the hope of an answer to the "overwhelming question" (see T.S. Eliot's poem, Prufrock). Moreover in this play the idea of a Christian God and 'salvation' is associated in the tramps' mind - at least in Vladimir's - with Godot. In spite of frequent allusions to Biblical and Christian sources none of Beckett's other works suggest the possibility of an answer - not to say look towards a solution- with a Christian connotation They tend to be secular and non-religious or even anti-religious, at least in the narrower sense of the term.

If WFG expresses the metaphysical anguish and the insufferable boredom of existence with its long wait for death, All That Fall (ATF) (Beckett 1965) is an_expression of the physical and emotional suffering and the triviality of existence. WFG is an abstract work, the characters are not realistic. ATF, on the other hand, presents recognizable, real life people of a small rural town presumably in Ireland. Mrs. Rooney, the main character, is a fat elderly woman, childless, having lost her daughter Minnie many years ago. She is walking with difficulty along a country road on her way to the railway station to meet her blind husband Dan. "What have I done to deserve all this, what, what?" she says as she halts. "How can I go on. I can't. Oh let me just flop down on the road like a big fat jelly out of a bowl and never move again!" (9). The loss of her daughter is constantly on her mind and she is full of self-pity. "Oh I am just a hysterical old hag I know, destroyed with sorrow and pining and gentility and church-going and fat and rheumatism and childlessness.....Minnie! Little Minnie" she wails for her lost child (9).

The play opens with a reference to Schubert's Death and the Maiden, the music Mrs. Rooney hears coming from a house as she walks past it. Soon a Mr. Tyler comes along riding his bicycle on the way to the station and greets Mrs. Rooney. Mrs. Rooney asks after his "poor daughter". She is fair, replies Mr. Tyler, but they "removed everything, you know, the whole...er...bag of tricks. Now I am grandchildless." Shortly a van passes by with "thunderous rattles" shaking Mr. Tyler up who gets off his bicycle just in time. "It is suicide to be abroad", says Mrs. Rooney, "but what is it to be at home? A lingering dissolution." (10-11). Soon Mr. Tyler is heard muttering something under his breath. When asked he explains, "Nothing Mrs. Rooney, nothing, I was merely cursing, under my breath, God and man....and the wet Saturday afternoon of my conception. My back tire has gone down again" (11). Beckett's plays are largely tragicomedies. The comic effect relies on exaggeration as well as the juxtaposition of the 'sublime' and the 'ridiculous', the philosophical or the serious remarks with the trivial and the particular. The comic element relieves the unbearable reality of existence as endured by the characters.

There is a great deal of material along these lines in the play, by way of reference to death, illness and other sufferings of life. The main event of the play is that the train bringing Dan Rooney to the railway station is late, the reason being that a child fell off the train and was crushed under its wheels. Ironically the theme of the sermon announced by the preacher for the next day is "The Lord upholdeth all that fall and raiseth up all those that be bowed down", the source of the play's title. Upon hearing the theme Mr. and Mrs. Rooney break into a "wild laughter". Clearly the child died a violent death and the Lord did not uphold it. The hollow pretensions of religion in the face of the cruel reality of the death of the child underline the irony of the situation. Life is contingent and cruel and there is no one out there to protect the innocent. What is more, the contingency and absurdity of the child's death is of the same order as the death earlier of a hen that is accidentally squashed under the wheels of a car when Mrs. Rooney was on her way to the station. "What a death! One minute picking happy at the dung, on the road, in the sun, with now and then a dust bath, and then – bang! – all her troubles over....All the laying and the hatching....Just one great squawk and then...peace." (16).

His play <u>Endgame</u> (Beckett 1964) consists of the cruel and overbearing character Hamm, his parents Nagg and Nell and his attendant (who may be his adopted son) Clov. At one point in the play Hamm asks them to join him in praying to God. "Our father which art in heaven", begins Nagg but Hamm cuts him short wanting them to pray in silence. They soon give up with Hamm exclaiming "The bastard! He doesn't exist". "Not yet" quips Clov (38). The entire scene has a touch of parody but Hamm's outburst, mocking or otherwise, reeks of anger and disappointment at God's absence. Clearly there is no one out there to pray to. Here the situation is not dissimilar to that in ATF. This is in clear contrast to WFG with its many allusions to the Bible, Christ and other Christian beliefs and symbols as well as the quest for meaning and the hope of 'salvation'. Yet even in <u>Endgame,</u> Hamm, who wants to exterminate all life, shows interest in finding some meaning or significance to the whole thing. Thus Hamm to Clov, "We are not beginning to…to…mean something?" Clov, "Mean something! You and I mean something! (Brief laugh). Ah, that's a good one!" "Hamm: I wonder….we ourselves….(with emotion)….we ourselves….at certain moments….(vehemently). To think perhaps it won't all have been for nothing!"(27) Yet the dominant note of the play is that this farce of an existence should end. If in WFG the phrase 'nothing to be done' conveys the principal theme, In <u>Endgame</u> it is 'why this farce day after day?'

The play begins with Clov's statement, "Finished, it's finished, nearly finished, it must be nearly finished". Hamm to Clov later, "Have you not had enough? Clov: "Yes! (Pause) Of what? Hamm: "Of this….this… thing". Clov: "I always had. (Pause). Not you?"(13) And while Hamm thinks "it's time it ended" his attachment to existence, whether through habit or instinct persists. Thus Hamm , "And yet I hesitate, I hesitate to… to end. Yes, there it is….I hesitate to…..to end." In fact as Hamm remarks, "The end is in the beginning and yet you go on" (44). Clov to Hamm later, "Why this farce day after day?" Hamm: "Routine. One never knows" (26). Nell, Hamm's mother asks the same rhetorical question: "Why this farce day after day?"(18). Echoing the Buddha's question to one of his disciples, Hamm at one point asks Clov, "Did you ever have an instant of happiness?" "Not to my knowledge" says Clov (42).

Beckett's characters are often presented as having some disablement or the other presumably as a symbol of human impotence, dilapidation and suffering. Thus Hamm is blind and on a wheelchair while Clov is unable to sit down. As one critic puts it, 'Sensitivity to suffering is evident on nearly every page of Beckett's writings.....There is hardly one of Beckett's people who is not either crippled, blind, dumb, rheumatic' or suffering from some other disability or from a combination of them (Hamilton and Hamilton 1976, 31). In a London taxi once Beckett found three signs asking for aid: for the blind, the orphans, and war refugees. His comment was you don't have to go looking for distress. "It is screaming at you even in the taxis in London" (31). His point was that conditions that we think are exceptional are, nonetheless, very much a part of life. They don't cease to exist just because we chose to ignore them. Echoing the first Noble Truth of the Buddha and the world view of Schopenhauer, Beckett believes in the centrality of suffering which 'opens a window on the real' (Proust, 28). Neither the mind nor the senses can grasp the 'mess' of existence, and emotions cannot be trusted as guides through the chaos. Suffering is the one quality of experience about which man cannot be deceived. 'It is constantly present, and even his ignorant senses and confused mind cannot misrepresent it' (Hamilton and Hamilton 1976, 78). Indeed suffering 'is the one stable point of reference in the Beckettian universe' (78.).

Christianity, for Beckett, is a tale of human suffering. Christ's suffering and death represents human misery and suffering relieved only by death. There is either no god or if there is he must be a malevolent tyrant. Beckett looks for an explanation or a cause for man's suffering but finds no answer (37). Since existence inevitably brings untold suffering it must be a kind of punishment for being born. Thus birth is the original sin for which man is punished by suffering and death. And it is not a question of the amelioration of human condition and the alleviation of suffering. These practical humanistic measures cannot touch the basic conditions of existence (38). The 'universe as experienced by our consciousness is no less arbitrary, cruel, and senseless' than if it was created by a God who wishes to fill the world with pain.

Man's cruelty to man, especially evident on the part of those with privilege and power, is another fairly persistent theme in Beckett's works. In Endgame Hamm's treatment of his parents and that of Clov, his attendant, is nothing short of brutal and sadistic. He recalls his callous treatment of a poor, starving man with a child who begged him for help. He was then a rich land-owner or something of the sort. It was freezing cold, just before X-mas. The man wanted Hamm to take him and his child in his service. There were others who needed help and whom he could have helped but did not. Instead he told the man, "Use your head, can't you, use your head, you're on earth, there's no cure for that"! (37) Hamm's remark is reminiscent of Pozzo's outburst In WFG, "That's how it is on this bitch of an earth" (Beckett 1954, 121). Pozzo too was a tyrannical and cruel master to his menial Lucky whom he treated like a slave. There is an even more general statement in his novel How It Is (Beckett 1964) where human beings are seen as torturers and tortured in turn in an endless chain of 'intimacies and abandons'.

For Beckett then the futility of existence combined with the cruelty and suffering that it invariably entails condemns it to nullity. His writings express the revolt against 'the intolerable imprisonment of man within the determination of cause and effect, of beginning and ending'. To Beckett, the 'meaningless limitations and compulsions of birth and death, and the universe which imposes such conditions on man can never be accepted' (Robinson 1969, 32).

In <u>How It Is</u> the human condition as a whole is portrayed as a journey in the dark with no direction or goal. The narrator is crawling through mud and slime as he describes his journey. He comes across other crawlers in the mud acting in turn as torturers and victims. Here Beckett creates an 'eternal, dark and silent world where the hero is imprisoned in the mud and where countless millions of other men also lie tormented and tormenting' and which is infinite. The narrator differs from Beckett's earlier characters in that he feels little of their anguish, their quest for meaning and significance. Rather he has 'resigned himself to eternal futility'. He endures without hope, simply fulfilling 'the demands of the place where he finds himself'. (Robinson 1969, 217) He wonders about the laws of this world of muck and speculates on the underlying pattern or order. In the end he abandons this attempt and accepts the only known reality, i.e. of the mud and his voice. Crawling in the mud seems to represent the journey of the human race through the centuries, across a "vast stretch of time through abject abject ages each heroic seen from the next" (Sen 1970, 104).

<u>Against Procreation:</u> Given Beckett's attitude to existence it is not surprising that his characters often express strong anti-natal views.

For Beckett, as for Buddhism, birth is the gateway to the pain and suffering that existence brings to human beings, with the inevitable end in decay and death. Thus ironically it is birth that brings death in its train. For Beckett the two are closely linked. His notion of 'wombtomb' is expressed by such aphorisms as "birth was the death of him" (see Stewart 2009, 169). In WFG, Pozzo's oft-quoted remark, "They give birth astride of a grave, the light gleams an instant, then it's night once more" (Beckett 1954, 333) is a more poignant expression of the same idea. Elsewhere in WFG, Vladimir exclaims, "Astride of a grave and a difficult birth. Down in the hole, lingeringly, the gravedigger puts on the forceps" (339).

Since Beckett sees life – at any rate the conscious existence of man - as a form of punishment, the unforgivable 'sin' of being born, of being ejected into the meaningless phenomenon of life is nothing short of a disaster. Ergo all those who give birth are guilty. What they commit is little short of a crime. Beckett's characters are often full of resentment and disgust for their progenitors. Thus even such a hallowed figure as the mother is a target of such feelings. Thus Molloy, "I speak.... of her who brought me into the world, through the hole in her arse if my memory is correct. First taste of the shit".(Beckett 1959, 16,). Molloy hates his mother because 'she was the cause of his birth and hence all the consequent suffering' (Robinson 1969, 28). In <u>Endgame</u>, Hamm the principal character of the play, calls Nagg his father "Accursed Progenitor" and "Accursed fornicator." Nagg is the 'hated and unforgiven arbitrary author of his existence' (269). The following exchange between father and son reveals the absurd nature of procreation.

Hamm: Scoundrel! Why did you engender me?

Nagg: I didn't know.

Hamm: What? What didn't you know?

Nagg: That it'd be you. (Beckett 1964, 35)

As Benatar argued earlier, procreation violates the freedom of the unborn, one who is brought into being without consent. On the other hand the procreator does not know what attitude the progeny will have towards his existence. Viewed in its totality it turns out to be an absurd and irresponsible act which inflicts life upon an innocent. However Nagg adds a caveat, "if it hadn't been me it would have been someone else. But that's no excuse" (38), which underlines the sheer contingency and absurdity of both the act and its consequences.

In Endgame when Clov detects a flea Hamm is alarmed: "A flea? Are there still fleas?... .But humanity might start from there all over again! Catch him, for the love of God"(27). Clov returns with the insecticide. "Let him have it! "says Hamm. Recall that Hamm presides over a household consisting of himself, his attendant Clov and parents Nagg and Nell. They are confined to a room, with windows looking outside. But outside this room apparently all life has been eliminated. Hamm's alarm at the existence of a flea is therefore justified. For where there is life there is procreation which could start the disastrous process of evolution once again. Here Hamm echoes, in the context of a macabre but funny play, the more seriously articulated thought of Hartmann along these lines encountered earlier (see Ch.2 above).

Beckett's attitude to procreation is very similar to Schopenhauer's. Procreation not only brings in its wake suffering and death but also a continuation of pointless existence. It is a reprehensible act. There is little doubt that Schopenhauer was a major influence on Beckett. But there are other parallels and influences such as Manichean and early Christian viewpoints with which Beckett was, directly or indirectly, familiar (Stewart 2009, 177). According to a leading Early Christian father, Gregory of Nyssa, "the bodily procreation of children.....is more an embarking upon death than upon life....Corruption has its beginning in birth and those who refrain from procreation....bring about a cancellation of death by preventing it from advancing further" (173). In the Gospel of the Egyptians we find the same idea, viz. to procreate is to nourish death. "To abstain from procreation is to hasten the end of the world and so defeat death" (173). Saint Augustine, whose works were well-known to Beckett, thought along lines not dissimilar. In The City of God he wrote "death is perpetuated by propagation from the first man, and is without doubt the penalty of all who are born" and "lust in opposition to the spirit.....is the conflict that attends us from our birth. We bring with us, at our birth, the beginning of our death" (174).

In connection with his play <u>Krapp's Last Tape,</u> Beckett's notes refer to one of the three Manichean prohibitions, viz. that of marriage and sexual reproduction. For Mani, birth is the imprisonment of the "true light" of the spirit. Abstinence from sex prevents this from occurring (ibid.). Procreation, in other words, keeps the spirit mired in worldly entrapment. In Manichaeism, as in Schopenhauer's thought, the best situation is never to have been born and Beckett's essay on Proust appears to echo this belief. As we saw earlier (see Ch. 2), according to Schopenhauer the will-to-live finds its strongest expression in sexual reproduction which perpetuates the sufferings of existence. Procreation therefore must be rejected.

In Beckett's (1996) play Eleutheria, which was only published after his death, there is a Dr. Piouk, a most explicit and fervid exponent of anti-natalism. He would "ban reproduction", "perfect the condom and other devices and bring them into general use", and "establish teams of abortionists, controlled by the state". Among other things he would encourage homosexuality and practice it himself. (Stewart 2009, 177). Dr. Piouk makes it clear that he is not against sex as such providing it does not result in reproduction. And while new life is to be prevented, for existing lives euthanasia will be an option but not an obligation. Although put in a crude and extreme form all this seems to be broadly in line with anti-natal attitudes we explored earlier. Ironically however, Piouk does not follow upon his own plans. He wants a child "to entertain me in my leisure hours, which are forever becoming briefer and more desolate" and secondly "so that he can receive the torch from my hands, when they are no longer strong enough to carry it" (178). Here Beckett seems to hint at the problem of putting anti-natalism in practice, and the role of self-interest in treating procreation and new life as a means to one's own selfish ends. As Stewart comments, 'if even the most enthusiastic exponent of the end of reproduction cannot follow his own dictates, humanity must inevitably be condemned to continue its guilty, problematic existence' (178). Nonetheless Dr. Piouk acts as a mouthpiece of a radical solution to bring humanity and its sufferings to an end. Here Beckett's fictional writings show a close affinity with the philosophies and prescriptions of Schopenhauer, Hartmann and, above all,

Benatar. Benatar, it will be recalled, believes that despite the logic and strength of his arguments against procreation, baby making will go on unabated (see Ch.3). He rightly identifies parental self-interest as the main reason for procreation, which also seems to be the point Beckett is making here.

Man's Incurable Optimism: Beckett is scathing about what he calls 'our smug will to live' and 'our pernicious and incurable optimism' (Proust, 15). As we saw earlier, from a Brahmanic and Buddhist standpoint such attitudes are seen as a result of 'ignorance', which keeps us chained to existence with its interminable cycle of births and deaths. However very few of Beckett's characters display anything like 'optimism'. In some ways the tramps in WFG, with their hope that a certain Godot will come and 'save' them or at any rate provide them with the key to the inscrutable nature of man's existence, may be seen as being optimistic. Their refusal to commit suicide and to persist in waiting for Godot against all odds could be seen as an example of their 'incurable optimism'. But their persistence in the search for meaning or 'salvation' implies the refusal to accept existence and religious consolation at their face value. In this sense they are not optimists. What Beckett calls optimism is a part of our smug will to live, the central idea of Schopenhauer's philosophy. This is not what we find in the despair of Vladimir and Estragon and in their boredom with existence.

As we shall see below, Winnie, in the play Happy Days, is perhaps the only Beckettian character that shows the mundane everyday optimism which the vast majority of mankind displays and lives by. It is not so much ignorance on her part as a form of 'adaptation' to the sufferings of existence, a way of coping by denying its terrible reality, by putting a positive gloss on life (see Zapffe and Benatar in Ch.3). Common religiosity, occupation with the trivia of everyday living , following a routine, evoking memories of the past, longing for love, telling oneself stories and the like constitute her survival kit. Here we are reminded of Zapffe's point about the variety of means deployed by humans to forget the frightful reality of existence as well as that of Benatar's Pollyanna principle (see Ch. 3).

Happy Days, (Beckett 1966) opens with Winnie, a woman in her fifties, buried in a mound of earth. Her husband Willie, almost a silent companion, is holed up in the mound behind her. The environment is one of "blazing light" with scorched grass around. It is very hot. Winnie speaks of "this hellish sun... this blaze of hellish light". In fact there is no day and night any more. A bell rings for Winnie to wake her up and another for her to sleep. The play consists almost entirely of Winnie's monologue, ostensibly with Willie as the listener.

The title "Happy Days" is largely ironic. The key phrase uttered again and again by Winnie is "That is what I find so wonderful". In Act One She is buried up to her waist and in Act Two up to her neck. She can barely move, has nothing to do, and finds it difficult to pass time. She longs for attention, indeed love, and is pathetically grateful to Willie for his occasional, monosyllabic responses. After one of these she says "Oh you are going to talk to me today, this is going to be a happy day.....another happy day" (19). And when he does not respond to her repeated calls she accepts the situation with resignation: "ah well...can't be helped....just one of those things…another of those old things….just can't be cured….poor dear Willie" (10)….no zest…for anything…no interest…in life" (11). Much of Willie's indifference to her and his lack of response she finds "very understandable…..Most understandable" (23). Willie sleeps most of the time and when awake reads a newspaper and spends time looking at pornographic pictures. In spite of all this, she dreams of Willie coming round to her side of the mound to "live this side where I could see you… I would be a different woman… Unrecognizable… But you can't, I know… What a curse, mobility!" (35) She thinks Willie still desires her. Towards the end of the play when he starts climbing up the mound and trying to approach her she thinks he might want to touch her face again and may be after a kiss. When he calls her "Win", she is delighted and finds it "another happy day!"(47)

The play shows that her way of coping is the common way – following a routine, relying on habit 'the great deadener', the 'guarantee of dull inviolability' and a source of security, as Beckett (1931, 19, 21) put it in his essay on Proust . She has a large handbag within reach which contains a miscellany of trivia, things which help her get through the day. She busies herself with such things as brushing her teeth, putting on her make-up, lipstick and all, scrutinizing her face in a hand mirror, brushing and combing her hair. She does these even though there is no one around and Willie can barely see her. As she says, "these things tide one over". Although she wishes she had Willie's "marvelous gift" for sleeping his time away she adds promptly, "can't complain….no no….mustn't complain…..so much to be thankful for…no pain…hardly any…wonderful thing that…many mercies…great mercies…prayers perhaps not for naught"

(12). She repeats the phrase "That is what I find so wonderful" often. Examining her toothbrush she finds the words genuine pure "hog's setae". She does not know what that means and finds out later from Willie. However insignificant she has found something new and that is enough. "That is what I find so wonderful, that not a day goes by....hardly a day...without some addition to one's knowledge however trifling.. .....provided one takes the pains" (16).

The play begins with the bell for Winnie to wake up. "Another heavenly day," she says gazing at the zenith. She prays and follows up with "For Jesus Christ sake Amen" and "World without end Amen"(9-10). She relies a great deal on words, on her endless chatter to carry her through the day. But she admits "even words fail at times....what is one to do until they come again? Brush and comb the hair...trim the nails ....these things tide one over" (20). In Act Two when she is buried up to her neck and cannot move any longer she still remains the same Winnie. As the bell rings she opens her eyes and greets the "day" with "Hail, holy light......Someone is looking at me still....Caring for me still...That is what I find so wonderful....Eyes on my eyes" (37). Despite her terrible state she persists in her belief in a benevolent god.

Although aware of the increasing heat she finds it "wonderful....the way man adapts himself....To changing conditions". Suddenly her parasol catches fire by itself. But this does not disturb her equanimity. With the sun "blazing so much fiercer down, and hourly fiercer, is it not natural things should go on fire....in this way...spontaneous like" (29). Then follows this harrowing remark, "Shall I myself not melt perhaps in the end, or burn....just little by little be charred to a black cinder" (29). However soon she consoles herself by claiming that nothing really has changed, her parasol will be there again next day to help her through the day, her mirror will be in her bag as usual. "That is what I find so wonderful, the way things…(voice breaks, head down)". She turns to her bag, brings out odds and ends, stuffs them back and finally brings out a musical-box, turns it on and listens to the waltz from The Merry Widow. She sways to the rhythm with a happy expression and forgets all about the heat and the fear of being charred to cinders.

At other times she simply recalls her past. "Oh the happy memories!... My first ball!...My second ball…(Long pause)…..My first kiss!... .A Mr. Johnson, or Johnstone…Very bushy moustache, very tawny…. Within a toolshed, though whose I cannot conceive" (15). Words and speech are important to Winnie as a way of passing time and forgetting her misery. She often quotes lines or bits from the classics, e.g. Shakespeare, Browning and others. These too she claims help to tide her over. "That is what I find so wonderful, that not a day goes by…..without some blessing…in disguise (20)". "What are those exquisite lines?... Go forget me why should something o'er that something shadow fling….(Pause. With a sigh.) One loses one's classics….not all…A part…A part remains…….That is what I find so wonderful, to help one through the day…..Oh yes, many mercies, many mercies" (43). But she is aware of the limitations of talking. "Stop talking now, Winnie….don't squander all your words for the day, stop talking and do something for a change.." (31). She starts filing her nail. "Keep yourself nice, Winnie, that's what I always say, come what may, keep yourself nice (32)".

However at times she is reminded of the grim reality of her situation and neither routine activity nor words can stop her being aware of it. But she puts a brave face on it and is prepared to "wait for the day to come…..the *happy day(* italics added) to come when flesh melts at so many degrees and the night of the moon has so many hundred hours….That is what I find so comforting when I lose heart and envy the brute beast" (16). The brute beast is to be envied presumably because it has no consciousness and above all no awareness of its inevitable death. For Winnie her day of release from earthly existence is also a "happy day" just as so many of her other days she considers to have been happy. She is resigned. Come what may she remains and will remain happy. She generalizes her attitudes and action enlisting "human nature" in her support. "There is so little one *can* do…. One does it all…. All one can… 'Tis only human…. Human nature…. Human weakness…. Natural weakness" (18-19). Elsewhere she says despairingly, "What is the alternative?..What is…"(17).

We should note, however, that for all her 'optimism' Winnie is not without the consciousness of suffering. She is not exactly a 'conventional' everywoman determined to look at the positive side of everything and carry on regardless. For surprisingly, along with the common bric-a-brac in her bag she also has a revolver. She brings it out accidentally while rummaging inside her bag and kisses it rapidly before putting it back (13). Later she brings it out again by chance and this time decides to leave it out by her side. "Oh I suppose it's a comfort to know you're there, but I'm tired of you…I will leave you out, that's what I'll do (26)". She is reminded of Willie's "Brownie" revolver. He wanted her to take it away from him in case he is tempted to put himself out of his misery (26). As with many other Beckett's characters, for Winnie and Willie too the possibility of suicide remains in the background.

Happy Days can be seen as a paradigm case of the "incurable" human optimism and the will to survive no matter at what price. This is underlined by the stark disparity between the terrible conditions of Winnie's existence – her own as well as of the environment she is in – and her resigned state and acceptance of it all. The play has an air of compassion and understanding for Winnie and her ways of coping with the dreadful exigencies of existence. Her love and longing for Willie, their limited but nonetheless real communication through which a degree of companionship is achieved suggests some redeeming features. Beckett here seems to be making some concession to human weakness, to the plight of human beings afflicted with conditions for which they bear no responsibility. It is not without interest that this optimism and will to live is expressed by a female character whereas Willie seems to have freed himself of the will to live and appears to have no interest whatsoever in existence. Willie is akin to other Beckett characters, males who have turned away from 'normal' existence altogether.

Concluding Remarks: The principal theme of Beckett's writings, especially as expressed in his plays, is man's gratuitous suffering during his absurd journey from birth to death. His writings reflect and echo the rejectionist worldview we explored in the chapters above. His themes have a close resemblance to those we came across in the religious and philosophical viewpoints considered earlier. Life as pointless and gratuitous suffering (Buddhism, Schopenhauer), the absurdity of conscious life with its need for meaning and the impossibility of finding one (Zapffe), the false promises and consolations of religion (Zapffe), the 'incurable optimism' of human beings in the face of their terrible fate and the variety of coping strategies (Zapffe, Benatar), birth as the gateway to suffering and death and the need to stop procreation (Schopenhauer, Zapffe, Benatar), the boredom of a pointless and meaningless existence (Schopenhauer) are all there in Beckett's work.

Beckett is not a philosopher but a creative writer. He has rarely expressed himself on what his writing is about, in short the 'meaning' of his plays and other works. It is only in his long essay on Proust, written early in his writing career, that we get some idea of his philosophical thought. There he speaks of how habit and routine deaden sensibility and protect humans from existential insecurity. He castigates man's 'incurable optimism' which sustains human beings and keeps the wheel of life turning. He quotes the Spanish poet Calderon approvingly about man's 'sin' of being born. And echoing Buddhism he puts suffering at the heart of human existence. Although one cannot read off a writer's worldview directly from his fiction there can be little doubt about the dominating presence, in his plays and other works, of the rejectionist themes we outlined earlier.

Jean-Paul Sartre: Contingency and Existence (Nausea)

The principal character of Sartre's novel, <u>Nausea</u> (1962), is Antoine Roquentin, a man aged about thirty. The novel is in the form of a diary in which Roquentin records the story of his experience and gradual discovery of the contingent nature of existence. He realizes that people and things are simply there for no rhyme or reason. This awareness of the superfluous nature of all existence, including his own, fills him with anguish and despair. What follows is the struggle to come to terms with his metaphysical predicament and the search for a personal solution. What makes it an existentialist novel is that it is about *living* this philosophical discovery and its implications as an individual – about feeling it as a personal crisis - rather than simply recognizing it as an abstract piece of knowledge.

Roquentin is a historian writing the biography of Marquis de Rollebon, a French diplomat and politician of the!8th century. He is in Bouville, a provincial city and port, where the papers and documents of Rollebon are deposited. The book begins with a strange experience that Roquentin has been having lately when he touches or holds objects. He picks up a stone on the seashore but has an unpleasant feeling in his hand and drops it immediately. When he holds the doorknob to enter his room at the hotel where he is staying he feels as though the doorknob has a presence and a personality of its own and recoils from it. There are similar experiences with other familiar objects and Roquentin finds all this quite unsettling. He cannot understand what is happening to him. It is only later that he comes to realize that for the first time in his life he is experiencing 'existence' as a sheer material presence of things bereft of their everyday, familiar, innocuous nature as useful objects, e.g. a doorknob, or with names to identify or classify them, e.g. a seagull. Like most people he took existence for granted. Never before did he have this feeling of encountering things – objects, living creatures including himself and other human beings – as sheer material presences.

At first he thinks it is some passing feeling. He writes "My odd feelings of the other week seem to me quite ridiculous today: I can no longer enter them" (Sartre 1962, 8). He hears the footsteps of the commercial traveler, who comes every week, coming up the stairs of the hotel. It "gave me quite a thrill, it was so reassuring: what is there to fear in such a regular world? I think I am cured" (9). However soon he realizes that something *has* happened to him. "I can't doubt it any more'. He is afraid of being in contact with objects. They give him a feeling of nausea. Like the other day when he picked up a pebble on the seashore he felt a "sort of nausea in the hands" (20). Another day when he entered the café at his hotel 'the nausea seized me, I dropped to a seat, I no longer knew where I was; I saw the colors spin slowly around me". And since then "the nausea has not left me, it holds me" (30). Other encounters and uncanny experiences of this kind follow and eventually, one day, Roquentin has something of a revelation while in the Municipal Park. At last he understands the source of his nausea and his feeling of the strangeness of objects, including human beings, around him. This is how he describes it.

Contingency: "I was in the park just now. The roots of the chestnut tree were sunk in the ground just under my bench. I couldn't remember it was a root. The words had vanished and with them the significance of things, their methods of use, and the feeble points of reference which men have traced on their surface" (170-1). What confronted him was simply the brute presence of a "black, knotty mass". Never before had he understood the meaning of existence. And now suddenly it revealed itself. 'It is there, all around us, in us, it

151

is *us'* and yet we do not see existing things as simply a dense, opaque, bewildering presence, without any rhyme or reason. "We were a heap of living creatures' with not 'the slightest reason to be there, none of us" (172), thinks Roquentin, yet we seem to be unaware of the superfluous, contingent nature of our existence. He comes to the conclusion that 'Every existing thing is born without reason, prolongs itself out of weakness and dies by chance' (180).

How does Roquentin feel about his discovery? "I can't say I feel relieved or satisfied; just the opposite, I am crushed. Only my goal is reached: I know what I wanted to know; I have understood all that has happened to me....The nausea has not left me ....but I no longer have to bear it as an illness or a passing fit: it is I" (170). It is this meaningless presence of things and living creatures, including himself, that gives him the feeling of nausea. "I was no longer in Bouville, I was nowhere, I was floating, I was not surprised, I knew it was the World, the naked World suddenly revealing itself, and I choked with rage at this absurd being. You couldn't even wonder where all that sprang from or how it was that a world came into existence, rather than nothingness. It didn't make sense......I stifled at the depths of this immense weariness" (180). And then suddenly "the park emptied as through a great hole, the World disappeared as it had come, or else I woke up – in any case, I saw no more of it: nothing was left but the yellow earth around me". (181)

<u>Responses to Contingency:</u> Roquentin thinks that people sense the superfluity of their existence but try not to face up to the reality. There are many ways of evading it, many ways of imposing a necessity on existence, e.g. by inventing a creator or a causal being, asserting

one's 'right' to exist, as the elites tend to do. Surrounding themselves with family, professional involvement, civic leadership and the like they gloss over the fundamental absurdity of their existence. He finds ample proof of this when he visits the local museum.

The museum was full of portraits of the local worthies - the Bouville elites - idealized by the painter. "They had been painted very minutely; yet under the brush, their countenances had been stripped of the mysterious weaknesses of men's faces. Their faces, even the least powerful, were clear as porcelain". However at the entrance was a painting entitled "The Bachelor's Death", a gift of the State to the museum. The painting showed the bachelor lying on an unmade bed "naked to the waist, his body a little green, like that of a dead man... The disorder of the sheets and blankets attested to a long death agony....Near the wall a cat lapped milk indifferently" (113). This man had lived only for himself and by a well-deserved punishment no one had come to his bedside to close his eyes. The painting was, felt Roquentin, a warning to him. He could still retrace his steps and get back to the fold. Over a hundred portraits were hanging on the wall in the room he was about to enter. With the exception of a few young people who died prematurely and a Mother Superior, none had died childless or intestate, none without the last rites. "Their souls at peace that day as on other days, with God and the world, these men had slipped quietly into death, to claim their share of eternal life to which they had a right. For they had a right to everything: to life, to work, to wealth, to command, to respect, and, finally, to immortality" (114). Not only God was on their side but by inventing rights and duties these leaders of men had provided themselves with a rationale for their existence. The visitors to the museum were full of admiration and reverence for these men. Soon Roquentin has had enough. He turns back. "Farewell, beautiful lilies, elegant in your painted little sanctuaries, good-bye, lovely lilies, our pride and reason for existing, good-bye you bastards!" (129)

Roquentin realizes that there were other ways of shielding oneself from the truth about existence. His meeting with the Self-Taught Man revealed some of these. The Self-Taught Man was a bailiff's clerk, one of the few people in Bouville that Roquentin spoke to. He was interested in knowledge for its own sake and was often to be seen at the Bouville municipal library. He invited Roquentin for lunch one day when Roquentin remarked, laughing, "I was just thinking ….. that here we sit, all of us, eating and drinking to preserve our precious existence and really there is nothing, nothing, absolutely no reason for existing" (151). The Self-Taught Man became serious making an effort to understand him. He repeated slowly, "No reason for existing". "You mean life is without a goal? Isn't that what one might call pessimism?" (151) He tells Roquentin about a book he read by an American writer called "Is Life Worth Living?" Isn't that the question he was asking? That certainly wasn't the question Roquentin was asking. But he had no desire to explain. "His conclusion", the Self-Taught Man says, consolingly, "is in favor of voluntary optimism. Life has a meaning if we chose to give it one. One must act, throw one's self into some enterprise. Then, if one reflects, the die is already cast, one is pledged" (151.). What do you think of that Monsieur, asks the Self-Taught Man. Roquentin replies, "nothing", and thinks, "that is precisely the sort of lie' that many people tell themselves".

(152). All this echoes Zapffe's point about 'anchoring' and 'distraction', about ways of avoiding facing up to the void of existence.

The Self-Taught Man then comes up with another line of defense. During the First World War he was taken prisoner. The experience of facing a common fate with other prisoners, and their close physical proximity gave him a sense of a strong bond of solidarity with these men. Although he did not believe in God, in the internment camp he "learned to believe in men" (154). He became a socialist and a humanist. The Self-Taught Man reminded Roquentin of the variety of humanists that he had come across in Paris: the Communists, the Socialists, the Christian humanists and others. They were all lovers of humanity in general even as they were at each other's throats (158).

Roquentin's companion makes one last attempt to corner him. Why was he writing? Surely in order to be read by someone? When he does not get an answer he says "Perhaps you are a misanthrope?" Roquentin knows this is a trap, an attempt to label him. If he accepts the label he is "immediately turned around, reconstituted, overtaken". Humanism can absorb all sorts of attitudes including misanthropy. For it too has its place in the human concert. It is "only a dissonance necessary to the harmony of the whole" (160).

Quite apart from the 'right' to exist proclaimed by the elites and the love of mankind or humanism as expressed, for example, by the Self-Taught Man there was also the question of being taken in by appearances. For example, people walking along the seashore in Bouville look at the sea and wax lyrical about it. "What a lovely day, the sea is green, I like this dry cold better than the damp" etc. Poets! Thinks Roquentin, they only see the surface which is a thin film of green. What they don't see is the reality under the water. "The *true* sea is cold and black, full of animals; it crawls under this thin green film made to deceive human beings…(but they)… let themselves be taken in" (167-8).

Roquentin's disgust with existence and its superfluity does not leave him. This contingent presence of "tons and tons of existence", including his own, is stifling. It is the source of his nausea and there is no way out of it. Above all his thought, his *consciousness* about it all is particularly unsettling. "It's worse than the rest because I feel responsible and have complicity in it. For example, this sort of painful rumination: I *exist*, I am the one who keeps it up. I. The body lives by itself once it has begun. But thought – I am the one who continues it, unrolls it. If only I could keep myself from thinking", thinks Roquentin (135). Ah! "Will there never be an end to it?' But my thought is *me*. That's why I can't stop thinking. At this moment 'I am horrified at existing". But "I *am the one* who pulls myself away from the nothingness to which I aspire: the hatred, the disgust of existing" (136). He thinks of suicide. It will at least wipe off one superfluous life. Nonetheless his body will go on existing. The blood, the decomposing flesh, finally the bones that the earth will receive all that would be "*in the way*". His death would be in the way. "'I was *in the way* for eternity" (173) concludes Roquentin. In other words once you come into existence you are stuck with it and even death cannot get rid of it. Here Sartre expresses the fundamental contradiction of *conscious* existence - the chasm between consciousness and mind on the one hand, and one's bodily self on the other which belongs to nature and appears to consciousness as an alien presence.

Roquentin decides to abandon his biography of Rollebon. He finds no point in one existent trying to recreate the life of another. Moreover when Roquentin thinks of his own past he can find nothing firm or reliable but only vague memories. If he can hardly hold on to his own past, he muses, how can he understand and recreate another's? It would be more like writing a work of fiction. He decides to abandon his project but has no idea what to do with himself. He writes, "I am free: there is absolutely no more reason for living, all the ones I have tried have given way and I can't imagine any more of them. I am still fairly young, I still have enough strength to start again. But do I have to start again?" (209) however he decides to move to Paris and, as we shall see later, he leaves with a project of sorts in mind which gives him some hope.

Suffering: Roquentin is very much a thinker. He is introspective, self-centered and concerned primarily with metaphysical aspects of existence. He personally suffers the anguish of being a part of existence which is without any purpose or justification. But he is not unaware of the common sufferings of human beings that he encounters in Bouville.

He hears Lucie, the cleaning woman at his hotel, complaining "for the hundredth time" about her husband to the landlady. "She has an unhappy home life", Roquentin informs us. "Her husband does not beat her, is not unfaithful to her, but he drinks, he comes home drunk every evening. I'm sure he is burning his candle at both ends ...It gnaws at her...she is morose all day.....weary and sullen. I hear her humming, to keep herself from thinking" (20). "it's there", she says touching her throat, "it won't go down". I wonder if sometimes she doesn't wish she were free of this monstrous sorrow, of these mutterings which start as soon as she stops singing…… if she doesn't wish to suffer once and for all, to drown herself in despair"(21).

One night, while out strolling on the Boulevard Noir he notices two people. The woman was pulling the man by his sleeve. The man says, "you are going to shut your trap now, aren't you?" 'But the woman still keeps talking. He pushes her away roughly and leaves without looking back. Suddenly deep hoarse sounds come from her, tear at her and fill the whole street with extraordinary violence. "Charles, I beg you, you know what I told you? Charles, come back, I've had enough, I'm too miserable." "Suddenly I recognize her. It is Lucie, the charwoman…This burning flesh, this face shining with sorrow. I dare not offer her my support, but she must be able to call for it if need be" (105). It is Lucie but "transfigured, beside herself, suffering with a frenzied generosity….she opens her mouth, she is suffocating…I am afraid she will fall: she is too sickly to stand this unwonted sorrow. But she does not move, she seemed turned to stone….she should be taken by the arm, led back to the lights, in the midst of people: down there one cannot suffer so acutely" (106).

On another day Roquentin looks at the local newspaper. It reports "Sensational news. Little Lucienne's body has been found…..The criminal has fled. The child was raped and strangled (107)". They found her body, the fingers clawing at the mud. "Her body still exists, her flesh bleeding. But she no longer exists". Her body violated, "She felt this other flesh pushing into her own…Raped". Roquentin cannot stop thinking of her.

The final episode concerns the Self-Taught Man.. AS Roquentin comes to return his books to the library he sees the Self-Taught Man sitting at the table with two school boys near him. He makes timid advances towards one of the boys lightly stroking his hand and whispering to him. He is spotted doing so by the librarian, a little Corsican. A fat woman sitting at a table nearby was also watching. The Corsican came up stealthily from behind watching him. "I saw you", he shouted, 'drunk with fury', "I saw you this time…Don't think I am not wise to your little game…And this is going to cost you plenty…We have courts in France for people like you". (221-2)

The Self-Taught Man made a feeble protest but went on reading. It is as though he was not taking any notice of the Corsican. Meanwhile the two boys left. Egged on by the fat woman the Corsican resumed his violent diatribe. Suddenly he gave a little whine and crashed his fist against the Self-Taught Man's nose. "For a second I could only see his eyes, his magnificent eyes, wide with shame and horror above a sleeve and swarthy fist….his nose began pouring blood". "I am going", he said, as if to himself. The Corsican hit him again. The woman next to me turned pale, her eyes were gleaming. "Rotter", she said, "serves him right" (224).

"I caught up with the Self-Taught Man at the foot of the stairs", writes Roquentin. "I was annoyed, ashamed at his shame, I didn't know what to say to him. 'Come to the drugstore with me', I told him awkwardly. He didn't answer…..His mouth and cheek were smeared with blood. 'Come on', I said, taking him by the arm. He shuddered and pulled away violently. But you can't stay by yourself, someone has to wash your face and fix you up' said Roquentin. 'Let me go, I beg you, sir, let me go'. He was on the verge of hysterics: I let him go"

(225). All these instances confirm Roquentin's feeling that what he sees all around him is a sort of 'messy suffering'.

Boredom: For Roquentin boredom is a part of existence. We are prisoners of time, which has to be passed. But time is not easily passed. "indolent, arms dangling, I go to the window. The Building Yard, the Fence, the Old Station – the Old Station, the Fence, the Building Yard. I give such a big yawn that tears come into my eyes"(45). He frequents the cinema and sometimes just goes to have something to eat in order to 'pass the time'. Elsewhere he writes, " I'm bored that's all. From time to time I yawn so profoundly that tears roll down my cheek. It is a profound boredom, profound, the profound heart of existence, the very matter I am made of" (210). Roquentin knows that boredom is simply the awareness of our suspension in time. And if we do not fill up time with some activity or distraction we experience the fundamental vacuity of existence in the form of boredom. He sees a game of cards in progress at the café and thinks "they do it to pass the time, nothing more. But time is too large, it can't be filled up. Everything you plunge into it is stretched and disintegrates" (32).

The 'solution' to existence: Roquentin has abandoned his biography of Rollebon and is preparing to leave for Paris. He still does not know what he will do. He has private means and does not need to 'work'. But what is he going to do with his life, his existence, given to him, as he says, for 'nothing'. He ruminates on these and other matters. And then a solution comes to him from unexpected quarters.

During his stay in Bouville, the only thing that freed Roquentin temporarily from his nausea and awareness of existence and even gave him a feeling of happiness was a jazz record that he heard from time to time at the café at the hotel. It was an old rag time: "Some of these days you'll miss me honey" sung by a black woman. He found the music almost moving. As he comes to say good bye to the patronne at the hotel before leaving for Paris, Madeleine the waitress holds up his favorite record and offers to play it for one last time. The record begins. The music and the song seem to have a life of their own. As the record plays Roquentin feels that it cuts through existence, the drab, messy, formless suffering that seems to surround him. It creates a world of its own, it moves through another time. Suddenly Roquentin understands why the music affects him in the way it does (233-4). It was the tune, the melody – something that did not exist yet had a life, a presence, a reality of its own. Although it unveiled itself through existents – such as the record, the gramophone, the needle – *it* was beyond their reach. If I were to get up, thinks Roquentin, and rip the record in two I will not reach it because *It* does not *exist*. It is beyond, yet it *is*. And he too wanted to *be*. That "is the last word. At the bottom of all these attempts which seemed without bounds, I find the same desire again: to drive existence out of me, to rid the passing moments of their fat, to twist them, dry them, purify myself, harden myself, to give back at last the sharp precise sound of a saxophone note" (234). Roquentin realizes that that is the secret of the appeal of this rather insignificant piece of music. It was created by existents - perhaps a jew in New York who wrote the song and a negress who sang it – but their lives seem to him almost justified. So "the two of them are saved.... May be they thought they were lost irrevocably, drowned in existence" (236-7). Roquentin thinks of them with a tenderness that he finds moving. They had "washed themselves of the sin of existing. Not completely, of course, but as much as any man can".

This idea suddenly knocks him over. And he thinks he too could create something that will have internal coherence and order, a life of its own. It will have to be a book but not the kind of history book he was writing. It will have to be some other kind of work. A story, a novel perhaps, that will be "beautiful and hard as steel and make people ashamed of their existence" (237). It will be a means of transcending existence with its time-bound and perishable nature. The work will endure and indirectly confer some meaning to his life as its creator. As he remarks, "some of its clarity might fall on my past". And then because of it he might be able to think of his life "without repugnance" (238). With this resolution the book ends.

Nausea and existence: an assessment: Sartre's work differs from the philosophical and literary perspectives examined earlier in that its rejection of existence is almost exclusively metaphysical. It is the contingent nature of existence that troubles Roquentin most. Although as we noted, he is not insensitive to the suffering he comes across, it is not life's pain and suffering that makes him condemn existence. It is the encounter between the reasoning mind, his consciousness with its need for meaning and purpose and the irrational nature of existence that is the source of the feeling of absurdity and superfluity in _Nausea_. Anti-natalism is of course implicit in the novel but perhaps the only direct reference to it is Roquentin's remark, "people are fools enough to have children" (212).

Roquentin's solution to the problem of existence is typically an intellectual and aesthetic one. It is by using his existence as a *means* of creating a work of art – in this case a literary-philosophical one – which he hopes to 'justify' his existence and make it acceptable to himself. We are reminded of Zapffe's idea of 'sublimation', one of the ways of coming to terms with existence. Where Sartre shares common ground with Zapffe, and to a lesser extent Benatar, is in his perception of the ways in which people evade the superfluity and absurdity of existence. For example, they do it by underpinning it with the idea of a God or creator, by defining reality in terms of social roles and relationships and by taking a benign and surface view of things, e.g. of nature, while ignoring its deeper and ugly reality. He castigates humanism for its worship of man and pooh-poohs the Self-Taught Man's notion of giving life a meaning by making a voluntary commitment to values or a cause. Roquentin considers this kind of justification of life as a 'lie'.

In short, all these forms of refusal to face up to the fundamental nature of existence amount to 'bad faith' or inauthentic modes of existence. This is important in that, ironically, later on Sartre's existentialism will take precisely the approach suggested by the Self-Taught Man and Sartre (1948) would claim his philosophy to be a form of humanism. But in *Nausea* he is quite radical and uncompromising in his rejection of existence. This work is paradigmatic of the rejectionist viewpoint albeit largely from a metaphysical rather than moral standpoint. Through Roquentin's personal predicament and anguish we can *feel* and experience the problem of the contingency and futility of existence.

Chapter 5

Summary and Conclusions
Rejectionism:
From Philosophy to Practice

The common thread running through the religious, philosophical and literary perspectives presented in the chapters above is the rejection of existence. The reasons for saying no to existence vary somewhat as do the paths of liberation envisioned in these different narratives. Naturally the literary perspective does not prescribe a course of action but is primarily an expression of the problematic nature of existence. However what they have in common, although the emphasis varies, is the keen awareness of the pain and suffering that living creatures have to undergo and the need to end this suffering. A second and subsidiary theme is the pointlessness and futility of existence which renders the entire process, including the suffering involved, unnecessary. A corollary to all this is that the 'good' that life also contains can in no way be regarded as justifying the 'evil', with pain and suffering as the predominant features.

Rejectionism does not believe in the calculus of pain and pleasure not only because any such exercise is impossible given that there is no common unit of measurement but also because its moral condemnation of existence is based on the irremediable presence of 'evil' in the world. It follows that to endorse existence is to condone evil, indeed to invite evil, albeit unintentionally. It follows that those who support and endorse existence are responsible, even if indirectly, for the crimes of humanity. To summarize: the rejectionist viewpoint has a long history stretching over millennia and a core of basic beliefs. Justifiably then rejectionism may be identified as a distinct attitude to life, a more or less coherent worldview.

An important point to be made is that If rejection of existence involves value judgment, so does its acceptance. For human beings the acceptance of existence is as much an ideological stance as is its rejection. But we seem to be a long way from realizing this. Instead life is accepted as simply natural, the default position so to speak. The vast majority of people outside the developed world reproduce 'automatically', i.e. without any thought or conscious decision. In the absence of contraception It just happens as a byproduct of coitus and having children is considered as simply 'natural' and normal. Here humans behave no differently from animals. Put in its social and cultural context it can also be seen simply as 'conventional' behavior. In the less-developed world and among the poor, with little education and the struggle to survive, we can scarcely speak of natalist behavior in ideological terms.

But among the people of the developed world with higher standards of life and education, 'choice' is a reality in regard to such things as marriage, procreation, and the number and spacing of children. Here we have to speak in terms of following, consciously or otherwise, an ideology of procreation and the perpetuation of existence. For in this context we can no longer put forward the excuse of acting 'naturally' or traditionally. To do so would be to act in 'bad faith', to borrow an existential concept. It would be to evade responsibility for our act. Each person has the obligation to think for themselves and consider the nature and consequences of their action. For the point is that we have moved far along the path of development. Increasingly it is no longer the 'natural' that shapes our lives and conduct. Rationality and technology have together moved our lives far away from naturalistic behavior.

Not surprisingly 'why children' is a question that is being asked increasingly in the developed world and the answers are often confused. Respondents are frequently at a loss to find coherent reasons for their natalist behavior. As one commentator (Ventura 2007, 1) puts it, 'Americans ask every conceivable question about children and receive endless answers from the expert and not so expert....but one most basic question goes unasked and unanswered: what are children for?' The same writer states that one possible answer is that they are needed to carry on the species and to pass on and extend the human heritage. Apart from this the 'biggest societal function that children serve today is to spend money or to have money spent on them'(2). His conclusion? 'When raising a family is a choice rather than a necessity (as it used to be in pre-industrial, pre-modern societies), we are on uncharted territory without map or compass and it's no wonder so many become irretrievably lost' (2). Nicki DeFago (2005, 52) reports that on the rare occasion when people are asked why they became parents they are mostly 'flummoxed'. A popular reason proffered is the 'maternal instinct' or 'biological urge' (52). Many parents do not think of it as a choice, she writes. Strong, if silent and indirect social pressure, ensures conformity to what is considered 'right'. A corollary to this is that voluntary childlessness still remains taboo, a form of deviant behavior (9, 12-3).

We should note, however, as Benatar and others, e.g. Overall (2012), point out children have a wide variety of 'uses' or functions. Potential parents may not be conscious of these since having children is considered simply 'normal' or 'natural'. But that does not mean that these parental and other interests involved are not important. Let us remind ourselves of a few of these. Starting a family, i.e. having your 'own' (genetically) child and raising it, is one of the principal reasons for marriage. It provides the couple with a 'life' together and a bond. It also locks in the parents and children in a lifelong relationship which is unique. The child is dependent on the parent until it reaches adulthood. Furthermore, in old age and illness or other situations of dependency both sides feel a moral obligation, if not also an emotional attachment, to care for the other. The emotional bonding between the parent and child remains an important intrinsic element. The upshot of all this is that the childless are likely to miss out on these and to live with the deprivations and other negatives that ensue. On the other hand they are spared the many hardships and frustrations of raising a child. In sum there are costs and benefits of childlessness whether voluntary or otherwise. Rejectionists therefore also pay a price for their choice, the deprivations being felt more acutely in old age.

Parental interests apart, we are still a long way from realizing that whether or not to support human existence is a question of moral and metaphysical choice. The individual has the right, and a duty, to say yes or no. The point is that both positions are ideological. Natalism can no longer be treated as 'natural' behavior that requires no justification (Overall 2012, 2-4). Why in spite of all the sufferings of human existence we wish to perpetuate it demands a clear and rational rather than an incoherent or conventional response. It is interesting to note that these issues are beginning to be recognized as important. Thus in a recent work on the philosophy of procreation, Christine Overall (2), who is not an anti-natalist observes 'In contemporary Western culture....one needs to have reasons *not* to have children, but no reasons are required to have them'. Having children is the 'default' position and not having is 'what requires explanation and justification' (3). Indeed she argues that these 'implicit assumptions are.... the opposite of what they ought to be'. The burden of justification 'should rest primarily on those who choose to have children' because bringing a new and vulnerable human being into existence needs 'more careful justification and reasoning' than non-procreation (3). Children cannot simply be a means to serve parental or other, e.g., societal, interests and she emphasizes the ethical dimension involved in procreation. But her logic concerning procreation could be extended to include existence itself. We take it for granted but it too needs justification. In short to the ethical dimension of procreation we need to add the metaphysical dimension. To procreate is to endorse existence with all that is implied by that decision.

In other words quite apart from the question of pain and suffering that human existence entails there is also the question of its sheer contingency and pointlessness. Is it really necessary to endorse existence via procreation and thus prolong it? This too demands a reasoned answer. Programmed by nature and socialized by the collective, who demands conformity, we are required to play the 'game' of life. But as one of Beckett's characters puts it, "why this farce day after day?" Where is all this leading to? After all there is no purpose, no goal or destination for the human race except its own perpetuation. Yet the need, indeed the yearning, for some transcendental rationale, some higher meaning or significance to it all has been a characteristic of humans and has led humanity to invent all kinds of excuses and rationalizations for our being here, primary among them being religious. But even religions are hard put to explain human suffering and injustices. What had the Africans done to be turned into slaves and worked to death in the plantations of Americas? What had the six million Jews done to deserve their cruel fate in the hands of Hitler? True, for the ever-present evils of the world religions have invented explanations. An egregious example is the Hindu doctrine of 'karma', a theodicy which sees the individual's fate in this life to be the result of his conduct in his previous life or incarnation. On the other hand, it is to the credit of ancient Hinduism to have judged moksha or liberation from the perpetual cycle of births and deaths as the supreme good that humans can aspire to.

Modern rejectionism is of course based on secular beliefs and has no place for supernatural phenomena intervening or controlling the world. Rejectionists believe that contingency rules nature and nature cares not a whit about individuals. Accidents decide so many things that are of supreme importance for the individual. In a deeper sense the contingent nature of existence means that there is no purpose or meaning out there which justifies life. Of course it is possible to embrace contingency and give life a meaning or purpose that it does not possess intrinsically. Taking life as a given we can then proceed to endow human world with values such as liberty, justice, compassion among others. This is the meaning of the existential premise that existence precedes essence. As Sartre (1948, 28) writes, 'man surges up in the world - and defines himself afterwards'. In other words it is up to each of us as individuals to create or choose values and forms of action. Thus humans lift themselves up so to say with their bootstraps and can go wherever they choose to. Existentialists go on to elaborate upon the vertigo, the dizzying feeling of total freedom and inescapable responsibility that human beings face in confronting such an arbitrary and yet consequential choice.

To start with, then, individuals can choose to respond in any way they like to the contingency of existence. This of course includes the decision whether to accept existence or to reject it. However, as pointed earlier, the idea of rejecting existence itself scarcely features in existentialist thinking (see Ch.3 above). It is not an aspect of choice that is explored in formal existentialist philosophy, whether that of Sartre or Heidegger. Indeed as we mentioned in the Introduction, atheistic existentialism is also essentially a value-free perspective which emphasizes choice but refuses to discuss, not to say prescribe, the substance of this choice. However it does take human existence for granted and seems to proceed on that basis. Is that a premise that involves value judgment? It is not clear but is implied by the claim that existentialism is a form of humanism (Sartre 1948). Unlike Sartre and Heidegger it is the value-committed philosophy of Nietzsche, who is also considered an existentialist, that discusses the question of acceptance or rejection openly as it does sexuality and procreation. And of course Nietzsche comes down strongly in favor of saying yes to existence and its perpetuation. He believes in progress through evolution and advocates a vigorous affirmation of life[1].

Rejectionism, on the other hand, is a philosophy which says no to existence. What it shares with Sartrian or for that matter Kierkegaardian existentialism is the freedom and the importance of individual choice which must include one's fundamental evaluation of existence. Thus rejectionism may be seen as a form of applied existentialism. It is somewhat paradoxical that Sartre, the philosopher of freedom, says nothing about the fact that we begin our life in unfreedom (we do not choose to be born), and in turn impose the same unfreedom via procreation on others. Put simply, procreation involves the enslavement of another, something that deserved a commentary from Sartre, the existentialist, who above all extols freedom and autonomy of the individual. Be that as it may the fact remains that neither the human species as a collective nor individual humans have chosen existence but find themselves saddled with it with no more meaning or purpose to their lives than any other living thing. In light of this pointlessness of existence rejectionism finds a dual objection to the business of procreation. It conscripts sentient beings to a lifetime of vexations and sufferings which they might be spared. Second, it perpetuates the unnecessary and pointless game of existence, taking it for granted as 'natural' and/ or legitimizing it with all sorts of rationalizations.

With increasing secularization religion as a principal means of legitimizing existence appears to have been weakening and we may expect it to weaken further. Nonetheless it has shown considerable resilience and persistence in the face of advances in scientific knowledge and technologies which impinge in profound ways on the religious view of life. Religion has shown considerable ability to adapt to the changing conditions of modernization. This together with the myriad of functions that it performs for believers in sustaining them through life means that its longevity and influence is not to be underestimated (see e.g. Pollack and Olson 2008). Indeed adherence to some form of religion, no matter what, remains an important marker of social conformity. Atheism still remains somewhat taboo, and open declaration of being an atheist subject to social disapproval. Rejectionism, on the other hand, presupposes a secular view of existence.

Religion apart, the principal mode of legitimation in conditions of modernity seems to be the idea of 'progress'. We might call it a form of secular faith. If religion is one form of 'opium of the people' the idea of progress can be called another. It appeals to believers and non-believers alike. Put simply, the core idea of progress is that reason together with scientific method has given us a powerful tool for the advancement of knowledge and its application to both material and social spheres. It means steadily improving material, social and political welfare throughout the world and a limitless horizon of progress and new discoveries. The idea of progress and the reality of it is of course relatively recent in human history. It goes back to 18th and 19th centuries, if not the 17th (Bury 1960; Nisbet 1994: 171-2). True, in a way the entire evolutionary process could be looked upon as 'progress' i.e. the evolution of the human species through the ages with its physiological and psychological dimensions including the development of consciousness (Bronowsky 2011). To this we could add material and cultural development, viz. language, writing, production of economic surplus and more recently increasing productivity and material affluence. In the 21st century the application of science and technology and the spread of market economy have had spectacular results in improving living conditions. Nonetheless the notion of progress, both as a concept and reality, remains highly contentious.

Looking back at the various perspectives we have discussed, which span three millennia, we find very little by way of reference to progress. While we would not expect it to be a feature of Hindu or Buddhist metaphysical thought, with its supernatural beliefs in rebirth and in cyclical time, we would expect the other philosophies, namely those of Schopenhauer, Hartmann, Zapffe and Benatar to consider the idea of progress and its relevance to their thought. Schopenhauer makes only a passing reference to technological progress, e.g. the coming of the railways, and dismisses it as of little relevance in the context of the all-consuming will-to-life and its attendant misery and suffering. This is in line with Schopenhauer's rejection of history and societal development as irrelevant to his timeless philosophy of the will and its resulting implications. By contrast, Hartmann takes the issue of historical development more seriously and discusses the question of progress in some detail drawing on a wide range of examples. But in the end he too dismisses 'progress' as making little difference to his assessment of existence. He writes, 'However great the progress of mankind, it will never get rid of , or even only diminish, the greatest of sufferings – sickness, age, dependence on the will and power of others, want, and discontent'(Hartmann 1884, v.iii, 103) . As for material improvements, new generations get used to them quite quickly and do not 'feel' them as anything special. Their general effect is to multiply needs and wants and any thwarting of these leads to greater discontent.

In fact the more important perspective on progress from Hartmann's philosophical standpoint is that intellectual, educational and cultural advance leads to a higher level of awareness, in short heightened consciousness, in humans. The result of this deeper awareness is that more and more people will see through our bondage to the will, the primitive will-to-live which keeps us under illusion and makes us do its bidding. In Hartmann's historical approach to the problem of existence 'progress' is a necessary condition, an 'urgent... requirement' in order for reason to prevail over will (115). Among rejectionists he is unique in embracing progress, not because of its amelioration of human suffering but because of its emancipatory potential. Secularization frees man from the grip of religion - the ancient justifier of existence with its supra-mundane beliefs - and as civilization advances and deepens further it emancipates man from the illusion of progress. For material progress shows at the same time that there is no moral progress as technology enhances the destructive potential of man and exposes the fundamental evil of existence more clearly. In short human development carries within itself the seeds of its own demise. For Hartmann liberation from bondage to nature and material existence is man's ultimate destiny and the highest point of spiritual awareness and affirmation. Interestingly enough, here he connects with the ancient Hindu and Buddhist thought with its conception of man's ultimate emancipation from existence - the "eternal recurrence" of births and deaths - as the supreme goal.

As for Zapffe, it is clear that progress makes no difference to what he sees as the fundamentally flawed and contradictory nature of human existence. What 'progress' does is to invent new forms of distraction. Zapffe was also an early ecologist of sorts and drew attention to the adverse consequences of urbanization and industrialization for the environment.

David Benatar makes no reference to 'progress' as such given that his argument, in the manner of Schopenhauer, is ahistorical. No matter how well-fed, clothed and housed people are every life is sure to experience some if not a good deal of pain and suffering. Moreover we know that the other side of technological and scientific advance is that it enhances human capacity to inflict pain, suffering and destruction on an ever larger scale. As Benatar (2006, 91) shows, the carnage and destruction of the two world wars, the atrocities of the Nazi and communist regimes, as well as other wars and revolutions of the 20th century constitute a mind-numbing record of human suffering and man-made evil. It is also witness to what is euphemistically called 'man's inhumanity to man'. It is as though what happened was an aberration, a deviance from human nature. No doubt it is comforting to assume that these were 'unusual,' 'unnatural' happenings brought about by 'monsters' such as Hitler, Stalin, Pol Pot, to name but a few. Clearly the record of the 20th century is a big blow to the Panglossian view of human progress. We should not forget that the 19th century was a period of great expectations, of peace and progress in the future through secular enlightenment, economic development and free trade between nations (Bury 334-9, Nisbet 171-2, 330). The history of communism in particular exposes the yawning gap between man's hopes and aspirations on the one hand and the cruel realities of human behavior on the other. Yet at the beginning of the 21st century we can see 'man's incurable optimism', as Beckett would put it (see above), taking over once again.

<u>The paradox of progress:</u> Does it make sense to speak of the paradox of progress in the context of rejectionism? Quite apart from the evolution of human species and the development of consciousness which enables the latter to question and to negate existence there is a more specific and modern development which strengthens the idea of a paradox. First, there is the liberalization of attitudes and ideas, associated in part with secularization, which makes it possible to express unconventional views and to have a reasoned dialogue about issues once considered strictly taboo. There are many examples notably homosexuality, contraception and abortion, suicide, voluntary euthanasia and death with dignity. It was not so long ago that some of these forms of behavior and practice were deemed immoral, indeed seen as crimes which entailed severe punishment. Of course many of the above, as well as attitudes such as atheism and rejectionism, are

still considered as forms of deviance and entail social sanctions and condemnation. Even in the developed countries religious authority, especially Catholic Church, remains intransigent and influential and we have to think here in relative terms. Undoubtedly however, 'progress' in the sense of liberalization of social norms, greater tolerance of different lifestyles and greater freedom of expression, helps to further the dissemination and practice of rejectionism. In the less developed societies the collective, notably the family as a group remains important for both physical and economic security. Children are almost a necessity for individual survival. It is only with the development of advanced industrial democracies that these conditions begin to lose their salience. In the post-industrial and post-modern society the individual comes into her own. Individuals and couples can live relatively safely in both physical and economic sense. The 'welfare state' is an important contributor in this context.

Perhaps the most important is the development of safe contraception which, thanks to technological advance and secularization, has become widely available. It is this above all which makes it possible to satisfy the coital urge and sexual desire without requiring abstinence and the frustration of sexual needs. The sundering of reproduction from sexual intercourse is a major step in weakening the grip of nature thereby extending 'choice' and the freedom to act. One might consider this important technological development as a necessary condition for anti-natalism to achieve widespread acceptance and support. Thus progress in these two interrelated spheres, those of ideas and technology, facilitate the rejection of existence. In this sense human development may be seen as a phenomenon hoist on its own petard. Alternatively anti-natalism can be seen as a part of 'progress' which enlarges our freedom of choice. Liberalization in the realm of ideas, greater control over reproduction, higher levels of education, and global communications help to raise awareness of the ethical and metaphysical dimensions of existence. Greater variety of beliefs and lifestyles become acceptable.

Hartmann, it will be recalled, saw development as an inherent unfolding of that evolution whose ultimate aim is the ascendancy of reason over will, leading to the negation of existence. True, he thought in terms of emancipation at the collective level through a global rise in consciousness, leading to a common resolve to end existence. But ignoring for the moment his teleology and emphasis on the collective, the logic of his argument has a great deal of relevance at the individual level. 'Progress' enlarges the scope of choice, and individuals can more easily decide not to procreate. Moreover it is not a mere coincidence that the voluntary childless tend to be more highly educated compared to the natalist population. In any case, both the *acceptance* of existence by way of reproduction and its rejection via non-reproduction are choices available to human beings. Underlying each of these choices are moral and metaphysical values and their affirmation.

The principle underlying rejectionism is that it is wrong to subject a sentient being to pain and suffering if it can be avoided. And this can be accomplished by abstaining from procreation. Rejectionism does value the 'good' that life also contains. However it is not prepared to pay the cost of that good by way of the evils of existence. Acceptance of life, on the other hand, in so far as it goes beyond simply being a form of naturalistic or traditional – including religious – behavior implies an acceptance of the evils of existence as part of a package which also contains much that is good. And each of these choices has a variety of implications. For example if everyone accepts the rejectionist approach then that means the gradual extinction of the human race. This unintended consequence of individual action is a part of the logic of rejection. On the other hand in practice it seems highly unlikely that the large majority of the world's population, including those in advanced countries, will give up procreation any time soon.

Rejectionism is a creed likely to appeal to a small minority of population. This minority will probably grow with time but at the moment it is difficult to predict the future since we know very little about people's attitude towards existence and how it might evolve as social conformity loses its firm grip on the populace and there is greater awareness of the ethical and metaphysical issues involved in procreation. What is important is to establish rejectionism clearly as a philosophical perspective on existence – as one of the possible existential choices - and to facilitate its practice throughout the world. It should take its place as one secular belief system among others. Here we need to distinguish between voluntary childlessness a) motivated by pragmatic considerations, e.g. lifestyle or lack of interest in parenting, which does not or need not involve any particular 'world-view', and b) that based on philosophical principles, notably prevention of suffering to future beings. It is important to recognize that the latter entails rejecting existence as the source of suffering whereas the former does not. This is not to deny that the distinction is 'ideal typical' and in reality the two may overlap. Furthermore, childlessness itself – no matter for what reason –presents problems and issues that rejectionists share with the others. Nonetheless it is important to differentiate between pragmatic and philosophical reasons for voluntary childlessness which sometimes get conflated under the term 'anti-natalism'.

Rejectionism in Practice: It is instructive to compare religious and secular approaches relevant to rejectionism. World religions, notably Hinduism and Buddhism (which we have discussed in this book), as well as Christianity, make a distinction between lay and virtuoso religiosity. The Hindu ascetics, the Buddhist and Christian monks may be seen as religious virtuosi with their abstemious lifestyles and wholesale dedication to their spiritual objective. Sexual abstinence is a part of this mode of life. All three religions lay particular emphasis on this since sexuality forms the strongest bond to existence and its perpetuation. It is a part of materiality and earthly desires which constitute an impediment to achieving liberation from bondage to nature and existence. The laity, on the other hand, lives a regular life as a part of the mainstream of society while following the moral and other guidelines of their religion. One can think of this as a form of stratification, a division between the religious virtuosi and the masses. Historically a gender division has also been involved, to a greater or lesser extent, often excluding women from the ranks of the virtuosi and in any case subordinating them to men in the hierarchy of status and power.

The underlying assumption has been that the necessary 'knowledge' and enlightenment, the renunciation of the world, the disciplined life, in short the challenging task of liberation can be truly desired and achieved by men and then only a select few. By contrast the masses are expected to marry, have progeny and continue social existence. In short, the goal of transcendence and emancipation, e.g. Buddhist nirvana, is reserved for the select few while the many are condemned to dull conformity.

The situation is very different with a secular belief system such as modern rejectionism. It is universal in its application and thoroughly egalitarian in nature. Any thoughtful person, male or female, rich or poor, highly educated or otherwise, can be a rejectionist. Furthermore, the rejectionist is expected to lead a normal life in every respect except one, albeit of a consequential nature, i.e. non-procreation. If reproduction is considered as an integral component of a 'normal' life then in that respect the position of the rejectionist is akin to that of a religious monk or nun. However unlike the latter the rejectionist does not have to practice sexual abstinence or any other form of asceticism. Moreover her beliefs are based on reasoning and evidence and her values are also clear with compassion as the core. No supernatural beliefs or imaginary state of affairs, such as that claimed for the state of Buddhist nirvana, are involved. She is not seeking a state of bliss or beatitude for *herself*. She is not self-oriented but other-oriented, seeking to protect future persons from the ills of existence.

There is another important difference. Religious monkhood, notably Christian and Buddhist, tends to be an organized community. Hinduism, however, accepts a plurality of approaches such as belonging to some form of 'ashrams' or retreats run by a Guru or an association, as well as living as a wandering monk, i.e. as a 'sadhu' or 'sannyasi'. The laity usually show respect for the sannyasis and support their livelihood through material gifts and other forms of assistance. While the comparison between monks and rejectionists may be somewhat far-fetched we need to note the similarities. The monks or nuns share a set of beliefs and thus constitute a 'community' of believers. At least in this sense rejectionists may also be said to form a community, people who share a set of beliefs. Of course monks are a part of an 'organized' community. Rejectionists, however small their number, have the potential to form at least a network, an association of some sort worldwide. Here again technological progress is a facilitator. The internet provides a viable means of communication enabling the exchange of ideas and mutual support.

In fact there is already a fairly robust presence of anti-natalism - a philosophical approach which rejects procreation in order to prevent suffering- in the form of websites, blogs and online debate. An interesting hybrid – partly opposed to procreation on environmental grounds – is VHEMT or voluntary human extinction movement which has been in existence for many years as an internet-based association. However it appears to be run almost entirely by its founder and consists of a loose network of 'members'. The only condition of membership is not to procreate[3]. There are also a large number of websites which act as a source of moral support and networking for the 'childfree' or the voluntarily childless irrespective of their reason for non-procreation[4]. There is some overlap between the two, e.g. the 'moral childfree' website which is against procreation on grounds of prevention of suffering but has little concern with related philosophical issues. What we have in effect is a plethora of websites and blogs with one group centered on 'anti-natalism' and the other simply on the state of being 'childfree'. It appears that the latter caters mainly for those who have chosen childlessness on pragmatic grounds. On the other hand anti-natalism, i.e. non-procreation in order to prevent suffering, is an expression of modern rejectionism. However because of the wider connotation of the term anti-natalism, a variety of prefixes, e.g. 'philosophical', 'philanthropic', 'altruistic', and 'compassionate', have been used to identify this particular form of rejection of procreation which is aimed at the prevention of suffering. While rejectionists need to make a common cause with the 'childfree' i.e. voluntarily childless on pragmatic grounds, the distinctive identity of rejectionism also needs to be affirmed and strengthened. For although it shares many problems with the childfree, such as social stigma, and coping with aging it constitutes a distinctive moral and metaphysical world-view. In this regard it is comparable to say 'rational humanism'.

Should rejectionists seek to acknowledge that the core of beliefs and principled action which they share constitutes a valid basis for a community or fraternity? Such an acknowledgment demands the creation of a stronger identity, and a vehicle for mutual support. Thus far modern rejectionism has operated largely as an individual and private belief system, a personal philosophy and attitude to existence. And despite a good deal of 'blogging' and online debate around it – under the title of 'anti-natalism' - much of it tends to be pseudonymous. Some of the blogs are rather bizarre conveying the impression of an esoteric doctrine, something rather like a cult. In any case many people are unwilling to identify openly with rejectionist beliefs. This is an important point and has not received much attention from rejectionists. The 'anti-natalist' websites are largely concerned with debating the philosophical underpinnings of non-procreation with little attention to its social aspects which underline the problem of being open about these 'counter-intuitive' beliefs. This raises the problem of legitimizing rejectionist beliefs.

The question is whether things should be left as they are, considering that publicly known rejectionists are often academics, intellectuals, writers and artists who tend to value individualism and privacy. On the other hand it seems that they may be just the tip of an iceberg of many ordinary individuals who are rejectionists but about whom we know very little and who would welcome the fact that there may be thousands of people worldwide who think and perhaps also act as they do. It makes sense that those who share certain basic beliefs and values should acknowledge this commonality and come together. This could help to strengthen their beliefs and resolve and spread the ideals of rejectionism widely, perhaps converting more people to their beliefs.

The word 'conversion' suggests an ideological movement with its publicity, recruitment, organization, membership, newsletter etc. Many rejectionists may have a strong aversion to such an approach and may prefer to retain the privacy of their beliefs and action. But this is a matter worth arguing about. For what is certain is that it is not easy to be an open rejectionist while remaining a part of the mainstream society. The rejectionist stance is likely to meet with strong disapproval not only from society in general but perhaps also from one's own relatives and friends. It can leave the rejectionist not only isolated but also vulnerable. It is an attitude to life that needs legitimization and placed at par with not only other, e.g. lifestyle-based, forms of non-procreation but also with natalist and pro-existence attitudes.

Furthermore we need to know a great deal more about, and understand much better, people's attitudes towards their own life and towards existence more generally. For this we need surveys, interviews and other forms of investigation. For example questions such as, 'given the chance would you like to live your life all over again?', 'is it fair to bring children into the world and expose them to all the suffering that living involves?' could well form a part of such an inquiry. On the face of it such questions and inquiries may seem bizarre to ordinary people but one has to start somewhere. Asking such basic questions about existence could be an excellent way of raising awareness of the underlying philosophical issues. It may be necessary to start with those likely to empathize with or appreciate the nature of such an inquiry. In short, we need a sociology of rejectionism to complement its philosophy. Compared with the latter the former is seriously under-developed. An Institute or Association of rejectionists could both attract people from a wider intellectual background and facilitate further exploration and thus greater acceptance of rejectionism. It could be an important step in moving it from the fringes towards the mainstream and furthering the legitimacy of rejectionist beliefs. An internet-based journal or newsletter may be another useful step.

In this context we need to take note of another important

difference between religious and secular rejectionism. Religion confers the all-important legitimacy on the ideology and practice of monkhood and world-negation. The case of Buddhism is particularly interesting in this regard. As Ligotti (2012, 130), for example, points out the Buddha's teachings are nothing if not thoroughly 'pessimistic' and world-negating. Yet Buddhism has a general acceptance and legitimacy which secular 'pessimists' e.g. rejectionists, lack. The reasons for this are not difficult to find. For one, Buddhism as a religion for the masses, especially in its Mahayana version, plays down the 'pessimist' aspects of the religion and emphasizes compassion and right way of life. For another the ascetic way of life and celibacy is reserved for the select minority of monks associated with the spirituality and mystique of nirvana. Moreover Buddhism is not against procreation as far as the mass of believers are concerned.

By contrast, as a secular and democratic belief system modern rejectionism is very different. There are no transcendent beliefs associated with it and it is primarily anti-natal in its orientation. Be that as it may, the important question is what can be done to promote the legitimacy of rejectionism? The first step in this direction might be to establish the contours of this world-view as a form of existential philosophy. In this regard Benatar's work represents a major step forward and seems to have been a catalyst for its further development. It is also important, as mentioned above, to institutionalize anti-natalism as a belief system. This could then underline the fact that both pro-natal and anti-natal attitudes are at bottom philosophical in nature, and involve existential choices. Rejectionism could thus emerge as a viable form of existential philosophy.

Endnote – Chapter 5

1. For a brief introductory outline to Nietzsche's thought see e.g. Earnshaw (2006). For an insightful discussion of Nietzsche's attitude to rejectionist view of life see Neiman (2002, 203-27).

2. The case for increasing secularization and the persistence of religion in the modern world is made by various authors in Pollack and Olson (2011). However they focus more on church membership, attendance, belief in god and such general indicators of religiosity. They have little to say about attitudes and beliefs concerning specific issues, e.g. contraception and abortion, and their relationship to secularization. An interesting idea offered in this book is that of 'belief without belonging' and its opposite 'belonging without belief'. The latter category will apply, for example, to Catholics who practice forms of birth control forbidden by the Church or who may be pro-choice regarding abortion.

3. See www.vhemt.org

4. See Basten (2009, 15-18) for a list of Facebook and Web-based groups.

Chapter 6

FAQs about Rejectionism

Q. What is Rejectionism ?

A. It is a philosophical viewpoint that is opposed to existence. It finds life inherently and deeply flawed in a number of ways. First and foremost life inflicts an inordinate amount of pain and suffering; second, it is totally unnecessary in that it is without any goal or purpose as such except its own perpetuation. Third, human existence is particularly reprehensible in that it inflicts life *consciously* upon innocent sentient beings, viz. children, who have not asked to be brought here and are thus victimized by being conscripted to the unnecessary process of birth, death and rebirth. Butchering and eating animals and subjecting them to cruelties of all kinds is another feature of human existence. Rejectionism is about moral and metaphysical rejection of existence on these grounds. The main implication of modern, secular rejectionism is abstaining from procreation. Another name for rejectionism might therefore be philosophical anti-natalism.

Q. Why this new term? Surely anti-natalism covers what you are saying quite well?

A. True, anti-natalism is the expression used generally for being opposed to birth. But it is too broad a term. Thus national policies for limiting population can be described as anti-natalist. Individual decision to remain childless, for whatever reason, can be described as anti-natalist. The focus here is on the result or the effect which is to prevent birth. It tells us nothing about the *reasons* for being against birth. That is where rejectionism comes in. It underlines the point that in this case anti-natalist behavior is based on some philosophical principles and attitudes towards life in general. This is different from deciding not to procreate for personal or some other general reason, e.g. environmental.

Q. Rejectionism sounds like nihilism?

A. It may, but rejectionism is not nihilistic. Nihilism is a creed, if it can be so described, that believes in nothing. It espouses no values. Rejectionism, on the other hand, is above all motivated by compassion for all living things and their sufferings. It is essentially a moral standpoint. It seeks to *prevent* future people from unnecessary pain and suffering. Not having children does involve sacrifices and deprivations for the childless. Rejectionists are prepared to pay this price on account of their beliefs. Rejectionism is a secular, not a religious belief system but in many ways it is similar to Buddhism.

Q. At any rate it does sound like a pessimistic doctrine ?

A. Pessimism or optimism, like beauty, is often in the beholders' eye. It is judgment from a *relative* standpoint. Rejectionism can be seen as hopeful in that it holds out hope for freeing human beings from bondage to nature, the evil of existence and the immorality of procreation. From this viewpoint pessimism implies the opposite, i.e. the attitude that 'there is no alternative', 'make the most of it', 'look on the bright side, 'you can't turn the clock back' etc. or pass the buck to God or nature.

Sometimes optimism-pessimism refers to a person's general outlook or disposition, i.e. whether it is hopeful or otherwise. Or it may refer to judgment about a specific matter, e.g. whether you are bullish or bearish about the movement of the stock market, the unemployment situation etc. But these have nothing to do with rejectionism. Admittedly both rejectionism and its opposite, i.e. conformity or acceptance, imply value judgment. Conventional wisdom is definitely in favor of 'optimism'. But the overused metaphor of 'the glass is half-full or half-empty' would be misleading in this context. The glass is *full*. The question is full of *what?* You have to drink it but it may be full of trans fat, saturated fat and other harmful substances although it may have some good things too, e.g. protein, and might taste good, like a lot of fast food or rich dessert. Rejectionism is based on the idea that don't invite others to the party where the food or drink is contaminated. That is not pessimism but common decency.

Q. Can you reject existence and still lead a normal life?

A. What rejectionism requires is non-procreation. Apart from that you can be and do whatever you like. You don't have to be an ascetic or in any other ways deprive yourself of anything. True, childlessness presents emotional and other challenges including care and support in old age. But these problems are not unique to rejectionists. You don't reproduce but you can adopt children. You may be a rejectionist and be fond of children. Not having your own children you might develop a closer relationship with your nephews and nieces and your sibling. You may forge strong friendships. You can marry, live with a partner, have sexual relations. None of these are against rejectionism.

Although you would have preferred not to be born you accept your own existence as a given and live your life like anyone else. Rejectionism, you have to keep in mind, is about prevention, *preventing* future people to be born. But other than that you lead a 'normal' life. There are of course a growing number of people who are not rejectionists but prefer not to have children for other reasons. They seem to lead a normal life.

Q. Isn't the anti-existential and anti-natalist stance of rejectionism very different from the positive, spiritual dimension implicit in the concepts of Moksha and Nirvana?

A. Certainly there is a difference. These two religious notions of liberation are premised on faith and a supernatural belief system. Both believe in rebirth of the same 'self', 'soul' or entity in different physical forms – human or otherwise – through the ages. What they seek is release of the 'self' from the interminable cycle of births and deaths. Liberation from existence is the essential aim or goal. In this respect modern rejectionism, with its non-procreation and rejection of earthly existence, is very similar to the ideal of moksha and nirvana.

Note, however, that there is much debate about being able to attain moksha or liberation in this life ('Jivanmukti') and what that means for the one who has attained that state. The state of *nirvana* in this life is even more contentious. The debate seems to end in pointing to a state that cannot be described in words or by any other mundane expression. It is considered as beyond subject-object distinction and thus implies some form of a mystical state. The Buddha, it will be recalled, refused to elaborate on what exactly nirvana meant and discouraged speculation about its nature. Let us also note in passing that what happens to the 'soul' or 'self' after liberation, i.e. post mortem, remains something of a mystery in the case of both moksha and nirvana.

By contrast, there is nothing mystical or mysterious about modern Rejectionism. It is entirely secular and this-worldly in orientation. Above all the crucial difference between moksha and nirvana on the one hand and modern rejectionism on the other is that the latter aims at liberating *future* generations, not existing individuals from the bondage of existence. Of course the individual self that is liberated in the case of moksha or nirvana, will indirectly liberate its future progeny also. But moksha and nirvana are essentially about the existing individual, i.e. they are ego-centric, whereas rejectionism is other-oriented or altruistic.

Q. What are the core values of rejectionism?

A. The rejectionist philosophy is essentially about *compassion*. It is about a deep empathy with the suffering of all sentient beings, especially humans, and a desire to *prevent* avoidable suffering. There are other values, notably, *meaningfulness*. Existence lacks an inherent rationale, a purpose or a goal which could justify putting up with all the 'evil' it entails. Of course one can think of many ways of justifying existence. But ultimately it all boils down to its acceptance and continuation simply because we find ourselves saddled with it by chance. This contingency or lack of a *reason* for existence is a part of rejectionist belief. Thirdly, *freedom of choice* is another value. Procreation imposes existence on beings who have not chosen to be born. It amounts to a form of enslavement or conscription which is an immoral act on two counts: violating the autonomy of a potential being, and exposing them to pain and suffering. These are the moral and metaphysical values underlying Rejectionism.

Q. Nietzsche argued that we who are a part of life cannot sit on judgment over life as a whole. It is wrong to set up an 'ideal' against which to judge real life and reject it because it does not measure up to the ideal. What is your response to Nietzsche's challenge that 'life is not an argument'?

A. Nietzsche is one of the great philosophers of the 19th century, if not of all time, and is often labeled as an 'existentialist'. As a young man he discovered Schopenhauer by chance and was bowled over by the radical atheism and the rejectionist implications of his philosophy. Although deeply influenced by Schopenhauer at first, Nietzsche later reacted strongly against his teachings. He totally rejected what he saw as Schopenhauer's 'resignationism', i.e. his call for the abnegation of the will and the practice of asceticism. The idea of the 'death of God' and the coming decline, if not demise, of Christianity made Nietzsche fearful of the spread of nihilism. His clarion call was to affirm life, to embrace the world which is all we have, to celebrate life with all its pain and suffering treating it as a price we have to pay for all that is glorious and wonderful in life and in human civilization. Nietzsche was a believer in evolution, development and, yes, progress which he believed came about through the boldness and creativity of great individuals. This forward and upward march of humanity is what life is or should be about. For Nietzsche that is its 'meaning', if we must have one. Nietzsche's general standpoint may be described as 'life for life's sake', i.e. that life transcends logic, truth and morality.

While Nietzsche has every right to hold on to and espouse his beliefs and values there is no basis for his argument that we as 'insiders' cannot pass judgment on life and that there is no vantage point outside the human community – sub specie aeternitatis, so to speak –from which to judge life. Here he is quite simply wrong. For as conscious beings we have lost our innocence which other living beings still have. We cannot help being aware of what life means and what it does to people. If human beings cannot make judgments about life who can? Only a theist could argue that humans have no right to judge life as a whole. Of course Nietzsche claimed that nature knows neither reason nor morality. So why should we expect these from her or apply these to her since life is a part of nature? Indeed Nietzsche rejected morality as a form of cowardice, a weakness, as something that is almost life-denying, something that inhibits and weakens the vital life force. The 'superman' of the future would trample over morality. Nietzsche recognized the irremediable conflict between nature and reason and came down on the side of nature, i.e. life. This is an existential choice and the rejection of reason and morality involves value judgment. Thus his attitude of acceptance and endorsement of life is as much a matter of choice as is its rejection. As 'grown ups' expelled from the Garden of Eden, we cannot hide behind naturalism. We have a choice and indeed a *duty* to choose. We cannot evade our metaphysical, and we should add moral, responsibility. As creators of value we also have to live by those values even if, paradoxically, they entail going against existence itself.

Nietzsche's greatness lies, among other things, in his recognition of the problem of 'evil' (i.e. moral evil) which he acknowledged as being an intrinsic part of existence. He was also fully aware of the role of religion and its theodicy in this context. Indeed we could say that how to justify a secular existence, which includes evil as an irremediable presence, was the philosophical problem he set out to solve. That life was a heavy burden which we should bear for the 'greater good' is implicit in his idea of 'eternal recurrence', i.e. that you must be able to embrace life to the point where you are prepared to accept and welcome repeated return to life, exactly the one you have led this time round. Consciously or otherwise Nietzsche here inverts the belief of Eastern religions which see the interminable cycle of births and deaths as a kind of 'life imprisonment', from which they seek release. Nietzsche, on the other hand, offers you a kind of eternal life – immortality – indeed asks you to will it as a cheerful Sisyphus rolling the same rock up the same hill ad infinitum. No wonder he puts the idea of eternal recurrence in the mouth of a 'demon'.

Q. If everyone stops procreation that would mean the eventual end of human species. How can a species will its own extinction?

A. True, species other than the human cannot go against the 'instinct' of reproduction. The survival and continuation of the species is a nature-bound, unconscious process. Of course particular species do die out as a result of natural disasters, environmental destruction or through natural selection and evolution. It is only human beings that have this emancipatory potential and the ability, if not also the obligation, to assume moral and metaphysical responsibility for our acts and to choose whether to reproduce or not. It is an individual choice. Throughout the ages holy men, and also women, of different religious persuasion have practiced celibacy although for reasons other than those of rejectionism. The latter adds another reason for voluntary childlessness.

However, collectively the human species is most unlikely to commit voluntary euthanasia. True if all humanity is converted to rejectionism the end result would be the extinction of our species. But it is more likely that the human species will become extinct through other means. No doubt we have vested interests in preserving our species and it may seem a disloyal act to go against this interest. But you have to think of the rejectionist as a 'conscientious objector' in a war where you are expected to fight for your country, right or wrong.

Q. Surely if rejectionists find life so bad shouldn't they advocate and commit suicide?

A. None of the rejectionist philosophies we have presented above advocate suicide. There is a misconception about the implications of rejectionism. All the approaches we have examined condemn existence on moral and metaphysical grounds, broadly conceived. Where they differ is in advocating different paths to liberation but suicide is not one of them. In modern thought about existential problems, it was Albert Camus who brought suicide into prominence linking it directly with the question 'Is life worth living?' For him a negative answer to this question logically entails suicide. Camus's well-known remark in this regard is worth quoting here: 'There is but one truly serious philosophical problem and that is suicide. Judging whether life is or is not worth living amounts to answering the fundamental question of philosophy. All the rest.....comes afterwards'. To put it in this way is to oversimplify both the question and the answer. The issue of existence and our attitude to it cannot be reduced to a simple either/or question. It does not occur to Camus that the more important question might well be whether to procreate or not, i.e. to bring a new life into what he recognizes as an 'absurd' existence. Be that as it may, let us review briefly our rejectionist philosophies and their attitude to suicide.

Both Hinduism and Buddhism recommend *withdrawal* from the mainstream life, eschewing worldly goals and desires, practicing asceticism and leading a life of contemplation. This is a path to 'holiness' and to ensuring liberation from the wheel of life, i.e. attaining the supreme good of not being reborn. Here suicide does not feature at all because it will not attain this goal. It is the penance, the process of purification that the practitioner goes through, the freeing of the self from worldly cravings and one's immersion into a spiritual inner world that leads to emancipation.

In the case of Schopenhauer, the process is not dissimilar albeit the context is secular. Liberation from the will and from pain and suffering is sought through the renunciation of the will. The goal is to achieve freedom from the will-to-live while remaining in this world.

It is similar to the Christian attitude of 'in the world but not of it'. It involves withdrawal from all worldly desires and goals and the practice of asceticism of the severest kind in order to weaken the physical basis of willing in order to attain a state of 'willlessness'. Schopenhauer discusses suicide at length and rejects it as an act of willing, an escapist act. It is a form of egoism and selfishness, not the practice of willlessness but rather a final and desperate act of willing to escape the misery of living. This is not something that Schopenhauer's philosophy, in spite of its strong emphasis on life as suffering, would advocate.

Hartmann's approach is even further removed from the idea of suicide. He too in common with Schopenhauer holds that life's pain and sufferings far exceed its 'good' including pleasure. But does he therefore advocate suicide? No. What he looks forward to is humanity's *collective* resolve to transcend existence. Far from suicide, which for him attains nothing but the ending of a few lives, he is in favor of continuing normal living, including reproduction. This is necessary in order for civilization to develop further and to reach a level of consciousness where the mass of humankind sees the folly and futility of existence, of doing the bidding of the will and remaining nature's accomplices.

Unlike the philosophies discussed above Benatar's approach focuses on how to spare *future* lives the pain and suffering of existence. And his solution is non-procreation. Suicide does not enter in this equation because he is not concerned with the liberation of *existing* people. The latter, as adults, are free to do as they please with their *own* lives. True he repudiates existence as an 'evil' primarily because it inflicts considerable pain and suffering, felt keenly by sentient beings. He has personally faced the question from some irate readers of his book " If you feel so bad about life why don't you go kill yourself?" He provides a detailed and cogent response to this challenge emphasizing the big difference between a) not starting new lives and b) ending existing lives through self-slaughter. The former is a preventive and peaceful act, the latter an act of violence and aggression against a person, a form of murder, albeit in this case of oneself. Rejectionism does not advocate killing. Quite apart from this fundamental objection there is also the question of the instinct of self-preservation, the vested interest one develops in one's existence and the hurt and trauma suicide is bound to cause one's relatives and friends. To Benatar's basic argument we might add that the rejectionist has an additional reason to go on living and that is to further the cause of rejectionism. That said it also remains true that the rejectionist approach is not opposed to suicide if an adult chooses rationally to do so.

Finally, Zapffe who was in favor of the gradual phasing out of the human race by limiting procreation below replacement levels or not having a child at all (he chose to remain childless though married), says virtually nothing about suicide. It is not something that he considers as a solution to the problem of existence.

Q. Does Rejectionism recognize that the world also contains much that is good, e.g. love, beauty, creativity, art, music, great literature, scientific and other forms of knowledge? If humans disappear all these good things will disappear too?

A. Indeed rejectionists recognize and appreciate the 'good' that the world also contains. The disappearance of these things will certainly be a loss. But think of the heavy price we pay for the good things to come into being. The question really is whether you are prepared to pay the price in all the negativities that the world also contains. How would you balance for example the horror, the agony, the torture, the outrage of Auschwitz against the works of Shakespeare, Bach, Newton and Einstein? Rejectionists, unlike for example Nietzsche, are not prepared to accept the view that for the sake of the positives of life we should accept all the evil that the world also contains. After all, life is a natural phenomenon; it just *is*. And we can either accept it or reject it. The choice is ours. As for our cultural achievements, they only make sense in the context of a human community. If humans disappear then no one will be left to regret the loss of these good things. At bottom human existence is no different from that of other species except that we are capable of providing all kinds of rationalizations for it, making a 'necessity' out of what is an entirely contingent affair.

Q. Do Rejectionists welcome death as a release from the bondage of existence?

A. Not necessarily. For once you have come into existence and grow up as a human being you become part of a network of relationships. You may be very interested or involved in your profession or in furthering some cause, i.e. quite apart from that of rejectionism, such as human rights or animal rights. The fact that you do not procreate does not mean that you cannot be an active participant in life like any non-rejectionist.

Death means saying farewell to the many good things of life that you come to appreciate. You also know that your own death changes nothing except that your own awareness and experience of the world – its good and evil – come to an end. Of course we can imagine many people, rejectionists and non-rejectionists alike, looking forward to death as a release from the ravages of aging including physical and mental decline. However our human 'conspiracy' requires that we do not complain and 'grin and bear it', that we keep going as long as possible. With the prolongation of life and with the coming of many medical interventions and technologies that can keep us alive even though our quality of life may have deteriorated severely the question of the 'right to die' and 'death with dignity' is assuming greater importance. Physician-assisted suicide and voluntary euthanasia are likely to become major issues in the coming decades. How many old people, often physically or mentally incapacitated, secretly long for deliverance we shall never know. This too is a taboo subject and it would be 'bad manners' to admit to anything of the sort. Here again we come across one of the paradoxes of progress and the 'absurdity' of conscious existence.

Q. Is rejectionism a form of Existentialism? What is the connection?

A. Existentialism is a philosophy which focuses on human existence and especially on the individual as a living and acting being rather than on such questions as the fundamental nature of reality, epistemology, logic and the like. With some simplification it may be described as a philosophical and literary perspective that emphasizes the freedom and the responsibility of the individual to live his or her life in an 'authentic' manner. Existentialism can be theistic or atheistic. It is the latter that is more relevant to rejectionism. This brand of existentialism starts with the idea of the 'contingency' of human existence. Heidegger speaks of it as 'thrownness'. Thus we are thrown into the world. We don't know why we are here. Moreover we have no control over the gender, class, nationality etc. of our birth. There is no god who has ordained our being here. And there is no divine guidance about how we should live. In Sartre's words "existence precedes essence". As conscious existences we are left to make of our life what we would. To quote Sartre again, "We are condemned to be free". It is up to each of us to decide not only how we live and act in the world but also our attitude to existence itself. Whether to say yes to existence, whether to have children or not are issues on which we have to make up our minds as free individuals.

The injunction to live life in an 'authentic' manner means that we should not be conformists who simply 'go with the flow' or make excuses for our attitudes and behavior, shifting the responsibility to others or to something external such as human nature. That is to act in 'bad faith' or in an 'inauthentic' manner. Our attitude and behavior should be based on our *own* thinking and feeling in full awareness of our situation and we should assume full responsibility for our acts. For it is through our action that we give meaning and value to existence.

Rejectionism also implies freedom of choice and the need to go against conventional attitude to existence. It is broadly in line with the key ideas of atheistic existentialism which, however, tends to dramatize its philosophy of freedom, choice and responsibility surrounding it with such notions as anguish, anxiety, dread, and despair. They are all about the burden of freedom and responsibility – moral and metaphysical - in the face of the void. The main difference between existentialism and rejectionism is that the former is in many ways a value-neutral philosophy. It offers an analysis of the nature of human existence and an approach or orientation to life without offering any substantive values and attitudes concerning existence. Authenticity, its key value, is formal and has no substantive content. Against the charge that one can, for example, be a good existentialist while choosing to be an authentic anti-Semite, existentialists argue that in choosing freely we must also choose and ensure the freedom of the other. Thus values are introduced through, as it were the back door, into what is essentially a formal philosophy of action. However, some secular thinkers labeled as existentialists, take a more substantive and value-based approach. For example both Nietzsche and Camus, in different ways, come down in favor of saying yes to existence. Emile Cioran, on the other hand, takes a rejectionist approach. He condemns existence in the strongest possible terms and states that one of the few things that he can feel proud of is his refusal to procreate.

Rejectionism is a value-oriented perspective on life with an emphasis not so much on individual choice and responsibility, which it takes for granted, but on substantive arguments and value judgments in support of rejecting existence. Thus it may be seen as a form of applied existentialism, a substantive and judgmental approach to existence. Another important point to note is that existentialism focuses on *existing* people, taking existence as given and reflecting on man's being-in-the-world and interaction with others. Rejectionism, on the other hand, is more concerned with *prevention*, with preventing future possible people from coming into existence. It also presupposes freedom of choice and responsibility. Indeed we could say it goes further in its concern for freedom and autonomy in that it considers the act of procreation as a violation of the autonomy of the one who is brought into being. On the other hand it has none of the heroic notion of the lonely individual making his choices in a mood of anguish and despair which has been characteristic of at least some forms (Kierkegaardian and Sartrian) existentialism. True, late 20th century existentialism has acknowledged the importance of the social context or situation ('freedom is situated') of the individual. However, as we pointed out earlier, rejectionism has yet to come to terms with the implications of the social context in which rejectionist ideas and decisions must take place. Existentialists (as well as others), on the other hand, would do well to develop a philosophy of procreation, one of the most under-theorized existential issues.

References

Basten, S. (2009) <u>Voluntary Childlessness and being Childfree</u>, The Future of Human Reproduction, Working Paper #5, St. John's College Oxford and Vienna Institute of Demography, Oxford: Oxford University Press.

Beckett, S. (1931) <u>Proust</u>, London: Chatto and Windus.

_____ (1959) <u>Molloy, Malone Dies, The Unnamable</u>, London: John Calder.

_____ (1954) <u>Waiting for Godot</u>, (Bilingual edition: Centenary Publication), New York, Grove Press.

_____ (1964) <u>Endgame</u>, London: Faber and Faber.

_____ (1964a) <u>How It Is</u>, London: John Calder.

_____ (1965) <u>All That Fall</u>, London: Faber and Faber.

_____ (1966) <u>Happy Days</u>, London: Faber and Faber.

_____ (1996) <u>Eleutheria</u>, London: Faber and Faber.

Belshaw, C. (2007) 'Review of D. Benatar's "Better Never To Have Been" <u>Notre Dame Philosophical Reviews</u>, June 9, 2007 http://ndpr.nd.edu/review.cfm?id=9983, accessed on 10/8/12.

Benatar, D. (2004) 'Introduction' in D. Benatar (ed.) <u>Life, Death, and Meaning</u>, Lanham, Maryland: Rowman and Littlefield.

_____ (2006) <u>Better Never To Have Been</u>, Oxford: Oxford University Press.

Bloom, H. (2010) 'Introduction', H. Bloom (ed.) <u>Samuel Beckett</u>, New York: Chelsea House Publishers.

Bronowski, J. (1974) <u>The Ascent of Man</u>, Toronto: Little Brown.

Bury, J. B. (1960) <u>The Idea of Progress</u>, New York: Dover Publications.

Buttner, G. (2002) 'Schopenhauer's Recommendations to Beckett' in A. Moorjani and C. Veit (eds.) <u>Samuel Beckett Today, 11</u> (Endlessness in the Year 2000).

Camus, A. (1975) <u>The Myth of Sisyphus</u>, London: Penguin.

Cartwright, D. E. (2010) <u>Schopenhauer: A Biography</u>, New York: Cambridge University Press.

Crawford, J. (2010) <u>Confessions of an Antinatalist</u>, Charleston, WV: Nine-Banded Books.

Darnoi, D. N. K. (1967) <u>The Unconscious and Eduard Von Hartmann</u>, The Hague: Martinus Nijhoff.

Defago, N. (2005) <u>Childfree and Loving It!</u>, London: Fusion Press.

Dienstag, J. F. (2006) <u>Pessimism</u>, Princeton, N. J.: Princeton University Press.

Earnshaw, S. (2006) <u>Existentialism</u>, New York: Continuum.

Flynn, T. (2006) Existentialism, New York: Oxford University Press.

Foster, C. (1999) 'Ideas and Imagination' in C. Janaway (ed.) The Cambridge Companion to Schopenhauer, Cambridge: Cambridge University Press.

Fox, M. (1980) 'Schopenhauer on Death, Suicide and Self-Renunciation" in M. Fox (ed.) Schopenhauer: His Philosophical Achievement, Brighton: Harvester Press.

Hamilton, A. and K. (1976) Condemned to Life: The World of Samuel Beckett, Grand Rapids, Michigan: William B. Erdman's Publishing.

Hartmann, E. V. (1884) Philosophy of The Unconscious (Vols. I, ii and iii), New York: Macmillan.

Harvey, P. (1990) An Introduction to Buddhism, New York: Cambridge University Press.

Hayry, M. (2004) 'A rational cure for prereproductive stress syndrome' Journal of Medical Ethics,30(4), 377-8.

Herman, A. L. (1983) An Introduction to Buddhist Thought, Lanham: University Press of America.

_____ (1991) A Brief Introduction to Hinduism, Boulder, Colorado: Westview Press.

Honderich, T. (ed.) (2005) The Oxford Companion to Philosophy, Oxford: Oxford University Press.

Janaway, C. (1994) Schopenhauer, Oxford and New York: Oxford University Press.

_____ (1999) 'Introduction' in C. Janaway (ed.) The Cambridge Companion to Schopenhauer, Cambridge: Cambridge University Press.

_____ (1999a) 'Schopenhauer's Pessimism' in C. Janaway (ed.) The Cambridge Companion to Schopenhauer, Cambridge: Cambridge University press.

Keown, D. (1996) Buddhism, Oxford: Oxford University Press.

Kluback, W. and Finkenthal, M. (1997) The Temptations of Emile Cioran, New York: Peter Lang Publishing.

Koller, J. M. (1982) The Indian Way, New York: Macmillan.

Lad, A. K. (1967) A Comparative Study of the Concept of Liberation in Indian Philosophy, Chowk, Brahampura: Girdharilal Keshavdass.

Ligotti, T. (2010) The Conspiracy against the Human Race, New York: Hippocampus Press.

Magee, B. and British Broadcasting Corporation (B.B.C.) (1978) Men of Ideas, London: B.B.C.

Magee, B. (1988) The Great Philosophers, Oxford: Oxford University Press.

Max Muller, F. (ed.) (1995) The Upanishads (Part Two), Delhi: Motilal Banarsidas Publishers.

Murdoch, I. (1993) Metaphysics As A Guide To Morals, London: Penguin.

Navia, L. E. (1980) 'Reflections on Schopenhauer's Pessimism' in M. Fox (ed.) Schopenhauer: His Philosophical Achievement, Brighton: Harvester Press.

Neiman, S. (2002) Evil in Modern Thought, Princeton: Princeton University Press.

Nicholls, M. (1999) 'The influences of Eastern thought on Schopenhauer's doctrine of the Thing-In-Itself' in C. Janaway (ed.) The Cambridge Companion to Schopenhauer, Cambridge: Cambridge University Press.

Nietzsche, F. (1954) in Kaufman, W. (ed.) Portable Nietzsche. New York: The Viking Press.

Nisbet, R.A. (1994) History of the Idea of Progress, New Brunswick, N. J.: Transaction Publishers.

Overall, C. (2012) Why Have Children? The Ethical Debate, Cambridge, Mass.: MIT Press.

Parfit, D. (1984) Reasons and Persons, Oxford: Oxford University Press.

Pollack, D. and Olson, D. V. A. (eds.) (2008) The Role of Religion in Modern Societies, New York; Taylor and Francis (Routledge).

Rawls, J. (1971) A Theory of Justice, Cambridge, Mass.: Belknap Press of Harvard University Press.

Robinson, M. (1969) The Long Sonata of the Dead: A Study of Samuel Beckett, New York: Grove Press.

Ross, F.H. (1952) The Meaning of Life in Hinduism and Buddhism, London: Routledge & Kegan Paul.

Sartre, J-P. (1948) Existentialism and Humanism, London: Methuen.

_____ (1962) Nausea, London: Hamish Hamilton.

Schopenhauer, A. (1969) The World as Will and Representation (Vols. I and II) (translated by E.F.J. Payne), New York: Dover Publications.

_____ (1970) Essays and Aphorisms (selected and translated by R.J. Hollingdale), London: Penguin.

_____ (1977) The world as Will and Idea (Vols. I, II and III) (translated by R. B. Haldane and J. Kemp), New York: AMS Press.

_____ (1974) Parerga and Paralipomena Vol.2 (translated by E.F.J. Payne), Oxford: Clarendon Press.

Sen, S. (1970) Samuel Beckett, Calcutta: Firma K. L. Mukhopadhyay.

Singer, P. et al (ed.)(1993) Practical Ethics, Cambridge: Cambridge University Press.

Smilansky, S. (1995) 'Is There a Moral Obligation to Have Children?' Journal of Applied Philosophy, 12(1), pp 41-54.

Snelling, J. (1998) The Buddhist Handbook, Rochester, Vermont: Inner Traditions.

Srivastava, P (2006) 'Conceiving a Child is a Sin' http://dontconceive.blogspot.ca/, accessed on 10/23/2012.

Stewart, P. (2009) 'Sterile Reproduction: Beckett's Death of the Species and Fictional Regeneration', in S. Barfield, M. Feldman, and P. Tew (eds.) Beckett and Death, London and New York: Continuum International Publishing.

Tangenes, G. (2004) 'The View from Mount Zapffe', Philosophy Now, 45 (4) March/April.

Taylor, L. and Taylor, M. (2003) What Are Children For?, London: Short Books.

Tsanoff, R. (1931) The Nature of Evil, London: Macmillan.

Vallee-Poussin, L. De La (1917) The Way to Nirvana, Cambridge: Cambridge University Press.

Ventura, M. (2003) 'What Are Children For?', The Austin Chronicle, Oct. 3, http://www.austinchronicle.com/gyrobase/Issue/column?oid=oid%3A1 80079, accessed on 6/23/2007.

VHEMT (2012) http://en.wlkipedia.org/wiki/Voluntary Human Extinction Movement, accessed on 4/20/2012. See also http://vhemt.org.

Watts, M. (2001) Heidegger, London: Hodder and Stoughton.

Weber, M. (1963) The Sociology of Religion, Boston: Beacon Press.

_____ (1968) The Religion of India, Toronto: Collier-Macmillan Canada.

Zaehner, R. C. (1966) Hinduism, London: Oxford University Press.

Zapffe, P. W. (2004) 'The Last Messiah' (translated by G. R. Tangenes) Philosophy Now, (45), March/April. http://scratchpad.wikia.com/wiki/The Last Messiah, accessed on12/11/2011).

Lightning Source UK Ltd.
Milton Keynes UK
UKHW040744140219
337321UK00001B/297/P